'Remarkable.'

Times Literary Supplement on *Weird Maths*

'A glorious trip through some of the wilder regions of the mathematical landscape, explaining why they are important and useful, but mostly revelling in the sheer joy of the unexpected. Highly recommended!'

Ian Stewart on *Weird Maths*

'A wonderful new book… if you love journeying into imagined mathematical worlds and simply exploring, then [this book] is pure, unadulterated escapism… brilliant.'

New Scientist on *The Biggest Number in the World*

'We are taken on an amazing adventure… [with] witty humour and fascinating facts… a comprehensive read that I would struggle to find fault in and for anyone with a passion for maths, or a knack for numbers, I couldn't recommend it enough!'

Astronomy Ireland on *The Biggest Number in the World*

'Gripping… dazzling tales of madness and derring-do.'

Brian Clegg on *Mayday!*

'A hugely enjoyable roller-coaster aerial ride in the company of oddballs and heroes, complete with vertiginous frights, dances of death, lonely impulses of delight and acts of mindless masochism in the name of science.'

Jonathan Glancey on *Mayday!*

'A compelling landmark book that will take you on an exhilarating ride.'

Daily Express on *We Are Not Alone*

'Informative, thought-provoking, and entertaining.'

BBC Focus on *We Are Not Alone*

ALSO BY DAVID DARLING

The Biggest Number in the World

Weird Maths trilogy

Megacatastrophes!
Nine Strange Ways the World Could End

Equations of Eternity

and others

KA-BOOM!

The Science of Extremes

DAVID DARLING

ONEWORLD

A Oneworld Book

First published by Oneworld Publications in 2024

Copyright © David Darling 2024

The moral right of David Darling to be identified as the
Author of this work has been asserted by him in accordance
with the Copyright, Designs and Patents Act 1988

All rights reserved
Copyright under Berne Convention
A CIP record for this title is available from the British Library

ISBN 978-0-86154-803-3
eISBN 978-0-86154-804-0

Typeset by Tetragon, London
Printed and bound in Great Britain by Clays Ltd, Elcograf S.p.A.

Oneworld Publications
10 Bloomsbury Street
London WC1B 3SR
England

Stay up to date with the latest books,
special offers, and exclusive content from
Oneworld with our newsletter

Sign up on our website
oneworld-publications.com

MIX
Paper | Supporting
responsible forestry
FSC® C018072

Contents

TECHNOLOGY

THE NATURAL WORLD

Introduction

WE LIVE BETWEEN fire and ice, between immensity and the unimaginably small, between the vacuum of space and the dark heat and pressure of deep rock. We live in moderation because life such as ours demands it. Our eyes see only a narrow band of the spectrum and our ears hear only a certain range of sounds because these modest sensory windows best serve our chances of survival. In size, in terms of orders of magnitude, we occupy the middle ground between the smallest conceivable thing and the universe in its entirety.

But this book isn't about the moderate or the middling. It's an unabashed celebration of extremes. It asks: what's the brightest light on Earth, the coldest place in the universe, the blackest material ever made, the lowest sound? It probes the boundaries of size and speed, depth and density, and reveals the stickiest, sweetest, smelliest and most poisonous substances known to science. *Ka-boom!* looks at the limits of what's been achieved, and what's possible, in both the human and natural world. In doing so, it presents not just a roll call of remarkable facts but also an exploration of the science behind the outer limits of the real world.

We may be average in many ways but our curiosity and desire to explore are boundless. Even as children we ask:

1

Where does the universe end? How deep could you dig a hole? What was the biggest dinosaur? Sometimes there's a practical aim in pushing the envelope of what's possible. New ways to hold vast amounts of information, materials that can withstand higher and higher temperatures – everything from non-stick surfaces to the latest smart phones are the products of investigating extremes.

More scientific and technical records will fall in the years ahead as we grapple with climate change, pollution, food security and other existential threats. A chemical sponge has been developed that can absorb up to ninety times its own weight in spilled oil and then be squeezed out and used again. The highest sustained temperatures on Earth will eventually be used to generate vast amounts of clean energy. But we don't always need a reason to push the envelope of current possibilities or to enquire as to what lies beyond the known: it's in our nature to wonder what's over the next horizon. So, fasten your seat belt, open your mind and prepare for a ride to the edge of the achievable.

PHYSICS

CHAPTER 1

How Low Can You Go?

TRY TO SING the lowest note on a grand piano and you won't even come close. Now imagine being able to hit a note that's a whole keyboard's-width lower. American singer Tim Storms doesn't have to imagine it: he holds the Guinness World Record for the lowest voice of anyone on Earth.[1] On March 30, 2012, he produced a sound that was slightly more than seven octaves below the bottom note on a grand piano. The sound was so low – just 0.189 hertz (cycles per second) – that his vocal cords, twice as long as those of an average adult male, were vibrating just once every five seconds. That's well into the infrasound range, below about 20 hertz, which is undetectable to human hearing.

Although most musical instruments aren't designed to play outside our normal auditory range – for obvious reasons! – some can produce infrasonic notes. One of these is the octobass, a gigantic version of the double bass.

The first octobass, built in Paris around 1750 and now on display in the Musée de la Musique, has three strings and stands 3½ metres (11½ feet) tall. A system of levers and pedals, connected to metal clamps on the neck, enables the player to fret the required notes while bowing in the conventional way. Only four functioning octobasses exist in the world today and only one, owned by the Montreal Symphony

Orchestra, is ever used in performances.[2] The lowest open string on the Montreal instrument sounds the note A_0, with a frequency of 27.5 hertz, but an octobass at the Musical Instrument Museum in Phoenix, Arizona, fitted with modern wire-wound strings, is tuned to produce a lowest note of C_0 (16.4 hertz), in the infrasound range.

Large pipe organs can go down to C_{-1}, or 8 hertz, which may seem pointless. However, just because we can't hear something that low doesn't mean we can't be affected by it. In 2002, a live experiment called 'Soundless Music' was carried out to explore the psychological effects of infrasound.[3]

The octobass of the Montreal Symphony Orchestra.

Under the guise of a concert featuring a variety of electronic and deep bass sounds, an infrasound generator was incorporated into the mix. Afterwards, people in the audience were asked to describe what they experienced. Many reported feelings of anxiety and foreboding, along with cold and tingling sensations. In the setting of a church or cathedral, it isn't hard to see how the lowest notes from an organ could evoke a similar emotional response that might be taken to be the effect of a supernatural or spiritual presence.

French scientist Vladimir Gavreau became a pioneer of infrasonic research following his unexpected encounter with ultra-low sounds in 1957. He and his team of acoustical engineers were working in a large concrete building when the group began experiencing bouts of nausea, at first assumed to be due to chemical fumes or some pathogen in the air. Weeks of investigation revealed the true source of the problem: a loosely mounted low-speed motor. The team built special equipment to detect the vibrations from the motor and eventually tracked the cause of their nausea down to infrasound waves with a frequency of 7 hertz. These waves from the motor induced a resonance in the ductwork and structure of the building, which amplified the original sound and led to its unpleasant physiological effects. The discovery triggered a wave of research into building acoustics in the ultra-low-frequency regime. Today, it's routine in new architectural schemes to test for and eliminate any infrasonic resonances and use sound-proofing where needed along with materials with special sonic properties.

Although the normal limits of human hearing are roughly 20 to 20,000 hertz, depending a lot on the individual and their age, some animals can hear much lower (and higher) sounds than we can. With their giant ears and bodies it comes as no shock to learn that elephants are among the heavyweights of the low-frequency domain, able both to detect and produce sounds down to 16 or even 12 hertz. Because ultrasound can travel far without much attenuation, elephants can use it to communicate over long distances. This may explain how groups of elephants, several kilometres apart, are able to travel along parallel paths, change direction simultaneously, and move towards each other in order to meet. Baleen whales, such as the blue whale, take infrasound communication to an extreme. Their low-pitched vocalisations can be detected over areas as large as an ocean basin.

Less obviously, ferrets, goldfish and some types of bird can sense infrasound. Experiments with homing pigeons have shown that they can respond to frequencies as low as 0.05 hertz. It seems they can use infrasound, produced by natural events, and which reverberate off the land and atmosphere, in their navigation. Where the local terrain or temporary atmospheric conditions, such as a temperature inversion, interfere with infrasound transmission, the birds lose their sense of direction.[4]

Thunderstorms, avalanches, volcanoes, large ocean waves, earthquakes and geomagnetic storms are among the powerful phenomena in nature that generate infrasonic waves. These waves, travelling quickly through the Earth, can be picked up by animals in advance of an impending disaster, and even serve as a warning that something destructive is on its way.[5] The earliest reference to animal behaviour of this kind comes from Greece in 373 BCE, when rats, weasels, snakes and even centipedes were seen to flee their homes and head for safety ahead of a destructive earthquake. In China, in the winter of 1975, an earthquake forecast was made based partly on unusual animal activity. As a result, many people chose to sleep outside their homes and were thus spared when a large earthquake did strike shortly after.

On December 26, 2004, captive elephants at a tourist site near the coast in Thailand began trumpeting and wailing in the early morning for no obvious reason. They broke their chains and stampeded up a nearby hill, pursued by trainers awakened by the commotion. But then the trainers heard a far more terrifying sound: the crash of an enormous wave as it smashed onto the shore, overwhelming everything in its path. More than 200,000 people were killed that day by a tsunami that had been triggered by an undersea earthquake.

We're all familiar with the squeaky, high voices of people who've inhaled helium. But other gases produce the opposite effect. Breathing in pure oxygen instead of ordinary air will give you slightly lower tones than normal, but to enjoy speaking with a really deep, Morgan Freeman-like bass for a few seconds one option would be to suck in some sulphur hexafluoride. (Don't try it at home, though: this gas can irritate your throat and lungs.)

The key factor affecting the pitch of voice is the speed of sound in the gas that passes over your vocal cords. The vibrating cords set up oscillations in the vocal tract which include the fundamental, or lowest frequency, together with a series of harmonics, or multiples of the fundamental. The speed of sound in helium is about 972 metres per second – nearly three times greater than the speed of sound in ordinary air. Because speed is proportional to frequency, the result is that when helium fills the vocal tract the frequencies of the resonant harmonics increase several-fold and a much higher-sounding vocal pitch is produced. The opposite is true in the case of sulphur hexafluoride, in which the speed of sound is a mere 133 metres per second, well under half the equivalent speed in air.

The lowest sounds ever detected, however, come not from anywhere on Earth but from sources that lie far away in space.

At a distance of about 250 million light-years is the Perseus cluster of galaxies. It's one of the most massive known objects in the universe, containing thousands of galaxies immersed in a vast sea of multimillion-degree gas. Near its centre dwells the galaxy NGC 1275 – a brilliant source of radio waves and X-rays, powered by a supermassive black hole. The black hole blows bubbles in the charged gas surrounding it, which in turn causes ripples to spread outward through the hot, thin medium of the Perseus cluster. The ripples are visible in the X-ray region of the spectrum and are the equivalent of sound waves propagating through air. The time between each wave is a staggering 9.6 million years.[6] In musical terms that equates to a B flat fifty-seven octaves below middle C on a piano – a billion times lower than anything the human ear can detect.

CHAPTER 2

Slo-o-o-w

THE THREE-TOED SLOTH has a well-earned reputation for being one of the slowest animals on the planet. It's certainly the slowest-moving mammal, creeping – when it moves at all – through its treetop habitat at an average speed of about 4 metres a minute. The sloth's metabolism, fuelled by leisurely munching on leaves and twigs, is as pedestrian as its lifestyle. It takes around a month for a single leaf to pass through the four-chambered stomach and digestive tract, and the creature needs to defecate only once a week or so, at which time it expels about a third of its total body mass in faeces and urine.

Just about the only time a sloth moves quickly is when it falls, which is surprisingly often. About once a week on average, the creature loses its grip and plummets to the ground. It might drop as much as 30 metres, or roughly the height of a ten-storey building, and reach a speed of 24 metres per second at the point of impact. But sloths are tough and unflappable, and generally crawl back up into their arboreal home, none the worse for wear.

On a large scale, everything about the sloth is sluggish (except for their occasional unscheduled tumbles). But it's a different matter when we descend to the sub-microscopic level. As much as 70 per cent of a sloth's body is made up

11

of water in which the molecules are darting around at about 600 metres per second or 1,300 mph.

All the things around us are made of atoms or molecules that are moving quickly – vibrating fast in the case of solids or barrelling along freely at high speed in gases and liquids. Strange as it may seem, one of the best ways to slow the particles in a substance way down is to employ the fastest things in nature – photons, travelling at 300,000 kilometres per second. In 2021, researchers in a lab at the University of Colorado used laser beams to chill a group of yttrium monoxide molecules to the lowest temperature ever achieved and thereby almost halt their motion.[1] The process is done in stages, steadily isolating the coldest and therefore slowest molecules so that, in the end, the 1,200 left are at just a millionth of a degree above the lowest temperature possible – absolute zero. They move so slowly that it would take them about an hour to cross from one side of a room to the other.

Among the longest experiments ever carried out, and one that's still running, is also the most boring because hardly anything ever happens. It started in 1927 when Thomas Parnell, the first professor of physics at the University of Queensland in Brisbane, Australia, heated a sample of pitch (a derivative of tar) and poured it into a glass funnel with a sealed stem. Three years were allowed for the pitch to settle, then, in 1930, the sealed stem was cut. From that date on the pitch has slowly oozed out of the funnel – so slowly that, up to the present time, only nine drops have fallen. The last one detached itself in April 2014, and, for the first time, was captured on camera.

The experiment stands in a display cabinet in the foyer of the Department of Physics at the University of Queensland demonstrating for all to see the fact that pitch, though it feels like a solid and is brittle enough to smash with a hammer, is really a fluid of very high viscosity, about 100 billion times that of water. If you're patient, you could be among the lucky ones to witness the fall of the next drop: a live webcam is trained on the famous black goo night and day. A similar experiment, at Aberystwyth University in Wales, was recently found to have been running since 1914, predating the Queensland set-up by thirteen years. But its pitch is stiffer and, even after a century, has failed to bear fruit. In fact, it's only just entered the stem of its funnel and is unlikely to produce its first drip for at least another 1,200 years.

The University of Queensland pitch drop experiment in 2012.

If that seems mind-numbingly slow, it's nothing compared to some other processes in nature. Xenon-124 is a radioactive isotope of the element xenon, a rare and highly unreactive gas. The half-life of xenon-124 – the time taken for half of the atomic nuclei in a collection of the isotope to decay – is about a trillion times longer than the present age of the universe. It's the slowest process ever witnessed by direct observation.

You may be wondering how something that takes on average about 160 trillion years to happen could ever be detected. It came about as a by-product of the search for another elusive aspect of nature: dark matter. The XENON1T dark-matter detector lies beneath 1,400 metres of rock in the largest underground research facility in the world, in the Gran Sasso e Monti della Laga National Park, about 120 kilometres from Rome. The detector contains 3.2 metric tons of xenon, including a small amount of the Xe-124 isotope. Although in any sample of a radioactive substance the *average* time for a decay to happen is given by the half-life, nuclear decay is a random process and some decays happen much faster. In fact, over a period of a year, the team at XENON1T detected the energy released from the decay of 126 Xe-124 atoms, a measurement that allowed them to calculate the isotope's incredibly long half-life.[2]

Surely, nothing human-made could run as slowly as a process that makes even cosmic timescales seem fleeting. Enter Dutch engineer Daniel de Bruin who, to celebrate reaching the grand old age of 1 billion seconds (in his thirty-first year!), built a machine to represent the number googol, which is one followed by a hundred zeros, or 10^{100}. The machine consists of 100 interconnected gearwheels each with a ten-to-one reduction ratio.[3] For the final wheel to complete one rotation, the first wheel in the chain would have to turn around a googol number of times. Given that it manages about 1,000 spins per hour, a googol revolutions would take 10^{97} hours or roughly 10 billion trillion trillion trillion trillion trillion trillion trillion years.

De Bruin's machine will obviously never achieve its goal or anything like it, for a mountain of practical reasons. But there is one process that would take even longer to complete and might actually happen – if the universe itself survives that long. It involves some of the most extreme objects in the universe: black holes.

A popular image of a black hole is of a terrifying, bottomless pit into which anything that comes too near must inevitably plunge, never to return. And it's true that the kind of black holes known to exist, at the centres of galaxies and the remains of giant stars that have exploded, do swallow up any matter that crosses the point of no return – the so-called event horizon. But, according to theory, black holes aren't completely black. They give off what's known as Hawking radiation. Over time, this would cause them to evaporate and eventually disappear.

The rate of evaporation depends on the black hole's mass. A mini black hole, only the size of a proton, would disappear in a fraction of a second in a flash of gamma rays. But bigger black holes hang around much longer. A black hole with the mass of the Sun would take about 10^{64} years to evaporate. The universe itself, by comparison, is a mere 13.8 billion years old. A supermassive black hole weighing as much as 100 billion Suns, such as exist at the centre of some large galaxies, could endure for 2×10^{100} years. Finally, if the universe survives long enough, it might outlive the slowest process ever theorised: the evaporation of monstrous black holes formed from the collapse of entire superclusters of galaxies. These gloomy, longest-lived of cosmic objects would give up the last of their contents by Hawking radiation after an astonishing 10^{106}, or 10 billion trillion trillion trillion trillion trillion trillion trillion trillion trillion years.

CHAPTER 3

Brilliant

IN A CITY renowned for its bright lights, the Luxor Hotel and Casino in Las Vegas, Nevada, has the brightest of them all. From dusk to dawn, from atop a black pyramid shines the Luxor Sky Beam. On a clear night, it can be seen by airline passengers flying at cruising altitude over Los Angeles some 440 kilometres away.

In a lamp room 15 metres below the top of the Luxor pyramid, the light produced by thirty-nine separate 7,000-watt xenon bulbs is collected and focused by curved mirrors into a single beam that shoots vertically up into the sky. As well as the intense light, plenty of heat is generated and the temperature in the lamp room rises to about 150 °C when in operation.

Not surprisingly, the Sky Beam has proved to be a big attraction – and not just to human tourists. Every night, millions of moths and other flying bugs are drawn to the brilliant glow. In turn, swarms of bats arrive in the evening to feed on the all-you-can-eat insect smorgasbord, while the bats themselves fall prey to opportunistic night owls. Besides predators, the fauna of the Sky Beam face the danger of the light itself. Anything that wandered into the beam would be instantly blinded and possibly also cooked – the temperature near the base of the beam is as high as 260 °C.

The standard unit for measuring intensity of light is the candela. Its name is the Latin for 'candle' and its definition, though sounding abstruse, continues to be based on the amount of light that a traditional candle gives off. In 1979, scientists defined the candela as: 'the luminous intensity, in a given direction, of a source that emits monochromatic radiation of frequency 540×10^{12} hertz and has a radiant intensity in that direction of 1/683 watt per steradian'. The frequency of 540×10^{12} hertz (cycles per second) is a very human-centred choice. It corresponds to a certain hue of green to which, of all the colours of the rainbow, our eyes are most sensitive. The value for the radiant energy in the definition approximates that of a typical candle. So, although the language may seem abstruse, it refers to something simple and familiar: how bright an ordinary wax candle appears when we look at it.

You need a lot of candles to illuminate a room reasonably well. The intensity of light from a 100-watt bulb is the equivalent of more than a hundred of them – about 120 candela. The brightest commercially available flashlight puts out a blindingly bright 450,000 candela. But even this seems barely more impressive than a glow-worm compared with the Luxor Sky Beam. The light that spears from the top of the Las Vegas landmark has an intensity of 42.3 billion candela.

The most brilliant natural light we see regularly is the Sun. It's roughly as bright as a thousand 100-watt bulbs at a distance of 3 metres. But the Sun is 150 million kilometres away, so its total light output is tremendous. Astronomers gauge the brightness of objects in space using a quantity called magnitude, of which there are two main forms: apparent and absolute. Apparent magnitude indicates how bright something appears as seen from Earth; absolute magnitude measures how bright it really is.

Magnitude is a reverse logarithmic scale, so that the brighter an object, the lower its magnitude number. It's a scale that approximates the ancient system invented by the Greek astronomer Hipparchus in the second century BCE, in which he assigned the brightest stars a magnitude of one and the dimmest stars a value of six. Every jump of one in the modern magnitude scale corresponds to a change in brightness by a factor of $\sqrt[5]{100}$, or about 2.512. For example, a star of magnitude two shines 2.512 times brighter than a star of magnitude three. The brightest objects in the sky have negative apparent magnitudes: Sirius –1.46, Venus –4.2, the full Moon about –13. On this scale, the Sun looks mightily impressive with an apparent magnitude of –26.8, or 10 billion times brighter than Sirius, the brightest star in the night sky.

But, of course, the Sun is cheating because it's so close, relatively speaking. Absolute magnitude evens the playing field by insisting that we place objects to be compared at the same distance. It's defined as the apparent magnitude that an object would have if seen from a distance of 10 parsecs – an astronomical unit equal to about 33 light-years or 310 trillion kilometres. On this scale, the Sun scores much less impressively, with an absolute magnitude of +4.83 compared to Sirius's –1.33. In fact, every star you can see in the night sky with your unaided eye is in reality brighter than the Sun. (The far more numerous stars fainter than the Sun are revealed only through telescopes.)

Among the true stellar luminaries we can catch sight of at night are the blue supergiants Rigel, at a distance of 863 light-years, and Deneb, about 2,600 light-years away, which have absolute magnitudes of –7.8 and –8.4, respectively. Both are in the order of 100,000 times brighter than the Sun.

Several of the most luminous stars known inhabit the Tarantula Nebula – a huge region of intensely active star formation within the Large Magellanic Cloud, a satellite galaxy of the Milky Way. Close to the centre of the Tarantula is BAT99-98, a stellar behemoth 226 times more massive than the Sun and about 5 million times more luminous. Only one other star has been discovered to date that significantly outshines it.

In a galaxy, far, far away – 10.9 billion light-years, to be precise – resides a star with no official designation known only as Godzilla. Thanks to a cosmic quirk, called a gravitational lens, the light from this distant galaxy is greatly magnified so that astronomers can see details within it, which would otherwise be invisible.[1] Some of the features in Godzilla's spectrum resemble those seen in very large, luminous and unstable stars in our own galaxy, such as Eta Carinae, which are very near the end of their lives. Godzilla shines about 15 million times more brightly than the Sun, but perhaps not for much longer. Soon, astronomically speaking, it is doomed to explode as a supernova and, for a few days or weeks, outshine an entire galaxy.

Only a tiny proportion of stars ever blow up. Those that do have at least eight times the mass of the Sun. A supernova can briefly reach an absolute magnitude of −19, so that if one went off just over 30 light-years away it would shine fifteen million times brighter than Sirius, or about 500 times brighter than the full moon. The supernova of 1054 CE, the remnant of which we can see today as the Crab Nebula, was visible during the daytime even though it lay 6,500 light-years away.

The brightest supernova on record is named ASASSN-15lh, after the All Sky Automated Survey for SuperNovae (ASAS-SN) telescopic survey that found it. A member of the elite class of 'superluminous supernovas', it was first detected on June 14, 2015, and found to lie in a galaxy 3.8 billion light-years away in the southern constellation Indus.[2]

At its peak it was 570 billion times brighter than the Sun and twenty times brighter than the combined light output of the Milky Way galaxy. Put another way, in its death throes it gave off ten times more energy than the Sun will produce in its entire lifetime.

Some astronomers have questioned whether ASASSN-15lh was a supernova at all. The same uncertainty hangs over another astonishingly bright event known as PS1-10adi, which was detected in 2010 using the Panoramic Survey Telescope and Rapid Response System (Pan-STARRS) at Haleakalā Observatory, Hawaii. PS1-10adi could be another superluminous supernova or a different kind of traumatic stellar death – a 'tidal disruption event' – in which a star is torn to shreds by the intense gravitational field of a giant black hole at the centre of a galaxy. Either possibility excites astronomers. Either offers them a chance to watch one of the universe's most extravagantly energetic processes at work.

Here on Earth, there's obviously nothing remotely as luminous as an exploding star. But how much light is given off in total by an object, and the intensity of that light, are two completely different things. A supernova and a hydrogen bomb explosion, for instance, are both incredibly bright events. The supernova is incomparably more luminous but how do they compare in terms of the *intensity* of their light? Ignoring the instantly destructive effect of standing at the epicentre of an H-bomb blast, the amount of light falling on a person's retina at ground zero would be about the same as that from a supernova half a light-year away!

In theory, there's no limit to the brightness of a beam of light. Unlike in the case of electrons or other subatomic particles, photons can be stacked on top of one another without limit. The only problem is a technological one: how to concentrate large numbers of photons in one place at the same time. In 2017, physicists at the University of Nebraska–Lincoln broke new ground in this quest by firing an ultra-high-intensity laser known as DIOCLES at electrons suspended in helium. The laser produced the most dazzling light ever made on Earth – 10 million times brighter than the surface of the Sun. It also resulted in some extraordinary and unexpected effects.[3]

The aim of the Nebraska experiment was to study how photons from the laser scattered off single electrons. Scattering of light is what makes most of the world around us visible. In normal light, only single photons scatter off the electrons inside atoms. But the super-high brightness of DIOCLES means that hundreds of photons at a time can bounce off a single electron, making the scattered light more energetic because it has the combined energy of all the laser photons.

The effect of the Nebraska experiment was to create scattered X-rays with unique properties. So bright was the laser that it altered the angle, shape and wavelength of the scattered light. Effectively, this means that, when illuminated by light above a certain threshold of intensity, things start to appear differently.

What's more, the X-rays generated by the laser beam hitting the electrons were powerful but lasted an incredibly short length of time and were held within a narrow energy range. This could enable sensitive 3D X-ray medical scans to be made using ten times lower radiation doses, for tracking down elusive tumours. The super-short duration of the X-ray burst also makes it work like the fastest of all strobe lights, freezing any high-speed motion. This should prove invaluable in studying chemical reactions that are impossible to follow using conventional X-rays.

CHAPTER 4

Shhh

JOHN CAGE WAS an American avant-garde composer perhaps best known for the quietest piece of music ever written. His piano composition *4'33"* calls for the player to sit in silence for 273 seconds – this being the number of degrees below zero on the Celsius scale of absolute zero at which molecular motion stops.[1] As Cage points out: 'There is no such thing as empty space or empty time. There is always something to hear or something to see. In fact, try as we might to make a silence, we cannot.'

Cage's *4'33"* breaks traditional boundaries by shifting attention from the stage to the audience and even beyond the concert hall. The listener becomes aware of all sorts of sound, from the mundane to the profound, from the expected to the surprising, from the intimate to the cosmic – shifting in seats, leafing through programmes, breathing, a creaking door, passing traffic, a recaptured memory. Not everyone is convinced this is art. In his essay 'Nothing', Martin Gardner wrote: 'I have not heard *4'33"* performed, but friends who have tell me it is Cage's finest composition.'

Sounds are all around us, although we're not normally conscious of many of them. Quieter sounds may be lost in the mix of louder ones, and, as we saw in Chapter 1, human hearing is restricted to a narrow range of wavelengths, in the same way that we can see only within a small band of the electromagnetic spectrum.

Sound intensity is measured in decibels, a unit named after the inventor of the telephone, Alexander Graham Bell. Roughly speaking, a 1-decibel difference in loudness between two sounds is the smallest difference detectable by human hearing. Like the Richter scale used to measure earthquakes, the decibel scale is logarithmic. Doubling the intensity of sound equates to an increase of just over 3 decibels. In going from a faint whisper of 1 decibel to normal speech at about 60 decibels, there's roughly a million-fold jump in intensity.

The quietest sound we can hear is generally reckoned to be zero decibels. But zero decibels doesn't mean zero sound. There can be negative decibels, which apply to sounds that are even quieter than we can detect. In any case, 0 dB is only a rough guide to the threshold of hearing. Some people are sensitive to much fainter sounds and our hearing is more acute when we're young. Also, the sensitivity of our hearing varies across the complete frequency range, from about 20 hertz (cycles per second) to 20,000 hertz, which it's possible for humans to hear, reaching a peak between 2,000 and 5,000 hertz.

It's hard to find natural places on Earth that are completely quiet. Getting away from the noises of civilisation is one thing but there are still usually the sounds of birds or the wind to disturb the silence. A place without wind and birdsong has to be somewhere barren and sheltered, such as a volcanic crater. One candidate for the quietest place on Earth is Haleakalā crater on the Hawaiian island of Maui. The more-or-less constant sound level here has been measured at just 10 decibels – about the same volume as your own breathing.[2] The only place quieter would probably be underground, in a deep cave, assuming there were no subterranean movements of water or drips from the roof.

'In space no one can hear you scream' was the iconic tagline for the movie *Alien*. But the claim that space is the quietest place in the universe is debatable. To survive in the vacuum of space you need to wear a spacesuit and a helmet in which there is air and therefore sound. Take off the helmet and you'd be dead within seconds. Scientists, though, are a resourceful bunch and don't let such trivial matters as survival get in the way of a good experiment.

To test the proposition that in space a scream will go unheard, a graduate student from Brunel University, London, and the BBC Radio show *The Naked Scientist* teamed up to send a microphone and a speaker into the upper reaches of the atmosphere.[3] A global invitation was sent out for people to record and submit their best screams, a selection of which made it onto the mission. Among the chosen yells was 'Children! Come and clean your room!' by Noha, a mother from South Africa.

Up and up rose the 'Screaming Satellite', reaching a height of 33 kilometres, where the atmospheric pressure is only about 3/1000th that at ground level, before the balloon carrying it finally burst. Just before this happened, Noha's exhortations to her offspring had faded to a barely audible whisper at the limit of what the mic could detect.

Wherever there's some medium for it to travel through, sound can propagate. There's sound on Mars, for instance, because Mars has an atmosphere. Thanks to NASA's *Perseverance* rover, which carries two microphones, we've been able hear the roar of a Martian wind, the clicking of the rover's laser as it zaps nearby rocks, and the hum of the spacecraft's miniature helicopter as it hovers overhead. But the atmosphere is colder, much thinner, and different in composition from that of Earth.[4]

The Martian atmosphere is about 100 times less dense than the air we breathe, so that sounds, such as the human voice, would be much softer on the Red Planet. On Mars you'd have to be much closer to the source of a sound to hear it at the same volume as you would on Earth. The Martian atmosphere consists largely of carbon dioxide, which is a better acoustic absorber than the nitrogen that dominates our own atmosphere, so sounds on Mars would seem more muffled.

On Venus you wouldn't survive long enough to get any words out or hear anything. But assuming you had some way to survive the ridiculously high temperature, the sounds you could hear and make would again be unique to that world. Venus's atmosphere is thick and soupy with pressures similar to those a thousand metres down in the ocean. The higher density would make vocal cords vibrate slower so that we'd all speak with a deep bass voice.

Back on Earth are special facilities, called anechoic chambers, which absorb almost all sounds made inside them. They're used in a variety of acoustic experiments, such as testing new audio equipment or the direction of noise from industrial machinery. It was during a visit to one of these chambers, at Harvard University, that John Cage got the inspiration for his 4'33". He wrote that, while inside the chamber, 'I heard two sounds, one a high and one a low. When I described them to the engineer in charge, he informed me that the high one was my nervous system and the low one was my blood circulation.'

The walls, floor and ceiling of an anechoic chamber are designed, and made of materials, so as to absorb virtually all the sound energy falling on them. The only sounds a person inside can hear are those coming from their own bodies or voice that travel directly to their ears. The effect is extremely unsettling and few people can stand the experience for more than a few minutes. Some anechoic chambers are occasionally open to the public. But visitors are generally not allowed in for any significant length of time without supervision. The longest anyone has stuck it out in one of these dead rooms is about forty minutes. Well before that, however, most of us would find the sound of our own bones grinding, the louder and louder ringing in our ears, and our loss of spatial awareness (another effect of the lack of reverberation), unbearable. The ultimate soundless hell, sometimes described as the place 'where sound goes to die', is the world's most advanced anechoic chamber: Building 87 at Microsoft's research lab in Redmond, Washington.[5]

The quietest sound in the universe is a single 'phonon'. Just as photons are the tiny, indivisible packets of electromagnetic energy in which light comes, so phonons are the smallest possible units of acoustic energy. In various laboratories around the world, scientists are developing devices that let them detect and control individual phonons or quantized packets of sound energy.

The anechoic chamber in Microsoft's Building 87.

No one can hear individual phonons and, in any case, the phonons used in quantum experiments have frequencies many millions of times higher than what humans can sense. These laboratory phonons are created on a chip made of piezoelectric material, which means any motion on the surface generates a voltage. As phonons travel across the surface of the chip, they give rise to minute voltages picked up by a sensitive transducer, which serves as both microphone and a speaker.

The quietest sound we use to communicate among ourselves is a whisper. In the future, the near-silent whisperings of phonons, each representing a quantum binary digit or qubit, may form the basis of a new generation of high-speed computers. So-called 'phi-bits' are less sensitive to environmental conditions than fragile quantum bits stored electronically. It may seem like the stuff of science fiction but the most powerful computers ever built, capable of supporting everything from advanced AI to complex cryptography, could eventually be based on the softest sounds that nature has to offer.

CHAPTER 5

Up to Eleven

IN THE 1984 movie *This Is Spinal Tap*, fictional guitarist Nigel Tufnel shows off his new amplifier that goes 'up to eleven', which he believes makes it louder than amps that are numbered only to ten. After the film was released, and as a nod to its humour, several prominent musicians, including Eddie Van Halen, began using equipment that could be dialled up to eleven or more.

In my years of going to concerts as a student in the early 1970s it was common for touring bands to boast about how much power their sound systems generated. But long before there was rock, some classical composers scored music that was intended to be played at high volume. The 1813 premiere of Beethoven's fifteen-minute work *Wellington's Victory* was performed by 100 musicians. *New York Times* music critic Corinna da Fonseca-Wollheim described it as a 'sonic assault on the listener' and the 'beginning of a musical arms race for ever louder ... symphonic performance'.

Louder music was made possible in part by innovations in instruments: trumpets with valves, metal flutes, and so on. The piano, invented in the early eighteenth century, saw some of the most dramatic changes. Early pianos used a combination of brass strings (for the lowest notes) and iron strings. As time went on, composers such as Mozart,

Beethoven and Liszt demanded more and more of the instrument, especially in terms of range and volume, so that the sound could fill ever-expanding concert halls. In response, designers added more strings and increased their tension, which meant they had to strengthen the frame with the addition of an iron plate. The biggest change began in the mid-nineteenth century, when steel strings were introduced. By 1912, these had reached their modern form with a tensile strength three times greater than that of the iron wires in the earliest pianos.

In some cases, steel wire also replaced the gut that had previously been used in all string instruments. One of the chief beneficiaries was the guitar, which in its new louder, steel-string incarnation found its way into folk and country music and, eventually, jazz and rock 'n' roll.

Among the loudest instruments in the orchestra is the French horn. A study by researchers at the universities of Queensland and Sydney in 2013 found that up to a third of French horn players under the age of forty had some noise-induced hearing loss.[1]

The loudest sounds ever used in an outdoor orchestral performance are those produced by live artillery in Tchaikovsky's *1812 Overture* – a piece hated by the composer himself, who described it as 'very loud and noisy and completely without artistic merit'. Indoor performances of the work normally employ drums, pre-recorded bangs, or other non-explosive substitutes. However, live cannon have accompanied the *1812* on occasions in the Royal Albert Hall.

As mentioned in the last chapter, the intensity of sound is measured in decibels (dB). On this scale, the quietest sound most of us can hear is around 0 dB (although negative decibels are possible). Each time the decibel level jumps by ten, the intensity of the sound *multiplies* by ten: a 10 dB sound is ten times more intense than a 0 dB sound, a 20 dB sound is 100 times more intense, and so on.

The gentle rustle of leaves reaches a mere 20 dB, the hum of a refrigerator about 50 dB, normal conversation perhaps 60 dB, and city traffic around 80 dB. The roughly 20 dB jump from ordinary speech to busy street noise explains why it's hard to chat to someone by the side of a town-centre street full of passing cars and buses. Although a 20 dB increase corresponds to a 100-fold leap in sound intensity, to our ears it seems only about four times louder. The fact that each tenfold jump in intensity is *perceived* by us as roughly only a doubling in loudness is one of the contributing causes of sound-induced deafness: we don't realise the damage that, say, loud music in headphones or other persistent exposure to high noise levels is causing.

A motorcycle registers about 100 dB and a thunderclap around 120 dB. Louder than that and we enter an auditory realm where sounds can lead to hearing loss, depending on the intensity of the sound and length of exposure. There was a time when some rock bands vied for the title of world's loudest. Deep Purple held the record for a few years in the early 1970s, then the mantle passed to The Who for having registered 126 dB at a distance of 32 metres from the speakers during a concert at The Valley in London in 1976. But much louder still are some of the most extreme sounds made both by human hand and nature.

Still on the subject of sound systems, the most powerful in Europe is LEAF, the Large European Acoustic Facility, which is part of the European Space Research and Technology Centre (ESTEC) in the Netherlands.[2] Embedded in one of its walls is a set of sound horns that can generate noise up to 154 dB when nitrogen is fired through them. Steel-reinforced concrete walls safely contain its noise and the system can be turned on only when all doors into the sound chamber are sealed shut.

A jet airliner taking off can be thunderously loud if you're standing less than 100 metres away. But, strange as it may seem, the loudest aircraft ever built was propeller-driven. The XF-84H experimental plane, manufactured by Republic Aviation for the US Air Force, was powered by a turbine engine connected to *supersonic* propellers.[3] Because the blades on the prop travelled faster than the speed of sound, even when idling on the ground, they produced a continuous, deafening sonic boom that radiated sideways for hundreds of metres. The shock wave they sent out was powerful enough to knock a man over. So loud was the aircraft, nicknamed 'Thunderscreech', that even when stationary at idle thrust it could be heard 40 kilometres away and was notorious for inducing headaches and nausea among ground crews. Its estimated sound level of 200 dB was not far short of that of the mighty Saturn V as it launched the Apollo missions to the Moon.

The loudest sound ever produced by mankind is thought to have been that of the most powerful hydrogen bomb ever exploded – the Soviet AN602, also known as the Tsar Bomba. Tested in 1961, it was more than 3,000 times as powerful as the weapons dropped on Hiroshima and Nagasaki, equivalent to 50 megatons of TNT. It registered on seismometers around the world and created a 224-dB roar at ground zero.

One of the greatest noises on Earth in recent historical times was heard thousands of kilometres away from its source. On August 27, 1883 the volcanic island of Krakatoa, located between Java and Sumatra, blew up.[4] Shortly after, the sound waves from the colossal blast reached the island of Rodrigues in the Indian Ocean, 4,800 kilometres away, and sounded to residents like distant heavy artillery fire. Five kilometres away, the sound level is thought to have been between 189 and 202 dB, while at the source it was around 310 dB.

Only one other event in the past century or so may have been louder than Krakatoa. On June 30, 1908, just after seven in the morning, a man was hurled from his chair outside a trading post at Vanavara in Siberia. At the same time, an intense wave of heat made him think his shirt was on fire. Fifty miles away, near the Tunguska River, something had exploded with the violence of at least a 10-megaton bomb – the equivalent of 600 Hiroshimas. Eighty million trees were flattened almost in an instant and now lay pointing radially out from the centre of the blast. The resulting seismic shock-wave registered on equipment as far away as England.

The generally agreed theory for the Tunguska event is that it was caused by an asteroid – about 40 metres across – that entered Earth's atmosphere at a speed of about 54,000 kilometres per hour and broke apart at a height of roughly 8,500 metres. The sound intensity at the point of the explosion has been put at between 300 and 315 dB.

If you stood near a giant rocket as it took off or were unlucky enough to be within a few kilometres of an event like Krakatoa or Tunguska, the sound alone would blow out your eardrums, shatter your bones and rupture internal organs. More than 160 decibels at close range is probably not survivable. The threshold of pain is about 130 dB, a sound intensity that the Krakatoa explosion exceeded by a factor of 250 billion.

In human terms we can't experience, up close, sounds as loud as the biggest that nature or our technology can produce because they'd instantly kill us. But there's also a *physical* limit to loudness. The loudness of sound in air is dictated by how large the amplitude of sound waves is compared to the ambient air pressure. At 194 dB, a sound wave has a pressure deviation equal to the normal atmospheric pressure at sea-level. Essentially, at this intensity the sound waves create a complete vacuum between themselves so that any further increase would mean the sound gets 'clipped'.[5]

Anything above 194 dB isn't sound in the conventional sense of a series of compressions and rarefactions. The energy of the source starts to distort the entire wave, producing not bigger and bigger peaks and troughs of pressure but a shock wave. Instead of sounds passing through the air, the result is a series of pressurised bursts. This is one of the reasons that large rockets produce a crackling noise rather than a steady roar.

As we saw in Chapter 1, although sound can't travel through a vacuum this doesn't mean that acoustic phenomena don't exist in space. Even in the near-emptiness between galaxies there's a thin gruel of gas that can act as a medium for waves akin to those of sound on Earth. The lowest notes ever 'heard', as we learned in Chapter 1, are powered by a supermassive black hole in the Perseus galaxy cluster, 250 million light-years away. The ripples of this ultimate *basso profundo* are also, in a sense, the loudest sounds ever detected in the universe. They're generated by a source with the combined energy of 10 billion trillion trillion Krakatoas going off at once.

CHAPTER 6

Almost 0 K

IF YOU EVER visit Vostok Station in Antarctica, wrap up well. Russia's scientific outpost on the southernmost continent, 1,300 kilometres from the Geographic South Pole, has a reputation for being a tad chilly. The warmest anyone's ever known it here, in the height of summer, is −14 °C. In mid-winter, it's brutally cold. On July 21, 1983, the temperature plunged to −89.2 °C – the lowest ever reliably documented anywhere on Earth.[1]

Oddly enough, despite being frigid all year round, Vostok Station is one of the sunniest places on the planet. In December the Sun shines for an average of 22.9 hours a day and there are more hours of sunshine here per year than anywhere in South Africa, Australia or the Arabian Peninsula, even though several continuous months go by in winter when the Sun never climbs above the horizon.

Temperatures undoubtedly sometimes fall below the official record. According to an unconfirmed report, Vostok reached −91 °C on July 28, 1997. Even lower temperatures must occur higher up towards the summit of the East Antarctic Ice Sheet where Vostok is located, and if wind chill is factored in, the all-time recorded low is −129 °C set on August 24, 2005.

Elsewhere in the Solar System are places that make the bitterest Antarctic winter night seem balmy by comparison.

As you might expect, worlds that are farther from the Sun are generally colder. There are plenty of reasons why 'Mars ain't the kind of place to raise your kids.' In fact, it's cold as −153 °C at the poles. Planet-wide the average temperature is about −63 °C with occasional summer equatorial highs of around 20 °C.

Beyond Mars lie the giant worlds Jupiter, Saturn, Uranus and Neptune and their many moons, all glacially cold. Enceladus, the sixth largest moon of Saturn, is almost entirely covered in fresh white ice, making it one of the most reflective bodies in the Solar System. Its noon-time high of only −198 °C is well below what it would be if the surface absorbed more of the Sun's rays. Much farther out is Neptune's big moon Triton, which reflects so much of what little sunlight it receives that its surface temperature hovers permanently around −240 °C.

Surprisingly, there is a world very close to Earth that can get as cold as Triton. In 2009, NASA's Lunar Reconnaissance Orbiter found evidence that some deep craters at the Moon's south pole are among the coldest places in the Solar System.[2] These craters are 'doubly shadowed', which means they're shielded not only from direct sunlight but also from secondary sources of heat, such as solar radiation reflected off nearby illuminated areas. Doubly shadowed craters are deep enough that sunlight hasn't reached their floors for billions of years. In the Stygian depths of these lunar cold traps the temperature may be permanently as low as −248 °C.

Only one place in the Solar System may be colder. Far beyond the orbit of Pluto lies a huge, more-or-less spherical region of space that's home to many billions of small icy-rocky bodies. This is the Oort cloud, named after Dutch astronomer Jan Oort, who first proposed its existence in 1950. While Pluto never gets farther from the Sun than fifty times the Earth–Sun distance – fifty 'astronomical units' (AU) – the inner edge of the Oort starts at around 2,000 AU. The distance from the Sun to its outer edge is uncertain but could be as much as 100,000 AU, or more than a third of the distance to the Sun's nearest stellar neighbour. Bathed effectively only in the feeble glow of starlight, the many lonely objects making up the Oort cloud may be as cold as −268 °C.

In the vastness of the universe there are surely countless locations that are colder than anything the Solar System has to offer. Case in point: the Boomerang Nebula. Located 5,000 light-years away in the constellation Centaurus, this glowing mass of gas is made of material gradually cast off by a star, similar in mass to the Sun, at the end of its life as nuclear reactions in its core came to an end.

In 1995, astronomers Raghvendra Sahai and Lars-Åke Nyman used the 15-metre Swedish-ESO Submillimetre Telescope in Chile to take the Boomerang's temperature.[3] It came out to be the lowest ever measured in the natural world: −272 °C. This is colder than the average temperature of the universe today (−270.4 °C) based on measurements of the cosmic microwave background – the much-cooled remnant glow of radiation from the Big Bang. The Boomerang has been releasing gas at a speed of about 140 kilometres per second for the past 1,500 years – an outflow that has gradually refrigerated the nebula in a process akin to the evaporation of sweat from your skin.

Temperature is related to the movement of particles – atoms or molecules – that make up a substance: the slower the movement, the lower the temperature. The lowest possible temperature, when all molecular motion stops, is known as absolute zero and has the value −273.15 °C. The Kelvin scale, named after the Scottish physicist Lord Kelvin (William Thompson), starts from zero (0 K) at absolute zero.

As scientists came to appreciate that there was such a thing as an ultimate minimum temperature, they also began a quest to try to reach it in the laboratory. Early efforts were made by Michael Faraday, who, by 1845, had managed to liquefy many of the gases known at the time. Using a combination of high pressure and immersion in a bath of ether and dry ice, he reached a new record low of −130 °C. Faraday believed that some gases, such as oxygen, nitrogen and hydrogen, were 'permanent' and couldn't be liquefied. But within a few decades he was proved wrong. Under conditions of sufficiently high pressure and low temperature, even the so-called permanent gases would become liquid.

In 1877, Louis Cailletet in France and Raoul Pictet in Switzerland produced the first droplets of liquid air at −195 °C. Six years later, oxygen was liquefied at a temperature of −218 °C by Zygmunt Wróblewski and Karol Olszewski in Poland. The last two hold-outs were the lightest gases, hydrogen and helium. Scottish chemist and physicist James Dewar liquefied hydrogen in 1898, reaching a new low-temperature record of −252 °C. His rival, Dutch physicist Heike Kamerlingh Onnes produced the first liquid helium, in 1908, at a temperature of −269 °C, by means of several precooling stages and a process called the Hampson–Linde cycle. He then went one better and, by reducing the pressure of the liquid helium, achieved an even lower temperature of about 1.5 K. These were the lowest temperatures achieved on Earth at the time, and for his pioneering work Kamerlingh Onnes was awarded the 1913 Nobel Prize in Physics.

At very low temperatures, substances start to display remarkable properties not normally seen, including super-fluidity (flowing without friction) and superconductivity (allowing an electric current to flow without resistance). Within a fraction of a degree of absolute zero, matter may also undergo a weird transition into what's been called a fifth state of matter, in addition to solid, liquid, gas and plasma. In this new state, known as a Bose–Einstein condensate, the particles in the substance lose their individual identity and behave like one giant super-particle. One of the main reasons that scientists strive for lower and lower temperatures is to explore these strange phenomena in more detail.

To get within a whisker of absolute zero, physicists deploy a range of novel techniques and unusual environments. In 2014, researchers at the National Laboratory of Gran Sasso in Italy cooled a one-cubic-metre copper container down to 0.006 K (−273.144 °C) for fifteen days, setting a record for the lowest temperature ever achieved over such a large volume.[4] In 2018, an instrument called the Cold Atom Laboratory (CAL) was sent up to the International Space Station. CAL takes advantage of the microgravity conditions aboard the ISS to maintain a Bose–Einstein condensate at a temperature of about 100 billionths of a Kelvin (0.0000001 K) for up to ten seconds.[5] That's long enough to allow experiments on this novel state to investigate fundamental laws of physics at the quantum mechanical level.

The current world record for low temperatures was set in 2021 by a team of researchers at the University of Bremen, Germany.[6] First, they trapped a cloud of about 100,000 rubidium atoms inside a vacuum chamber, then used a technique called magnetic lensing to cool the chamber to two billionths of a degree above absolute zero – in itself a new record – and transform the rubidium gas into a Bose–Einstein condensate.

In the second phase of the experiment, the chamber was put at the top of the European Space Agency's drop tower at Bremen's microgravity research centre. As the chamber plunged down the 120 metres of the tower, in free fall, a magnetic field was switched rapidly on and off. When the field was on, the rubidium gas was made to contract; with it off, the gas expanded. The effect of this switching back and forth, in tandem with the microgravity condition, was to slow the rubidium atoms' motion to almost nothing. For about two seconds, the interior of the chamber was the coldest place in the known universe – a mere thirty-eight trillionths of a Kelvin above 0 K.

CHAPTER 7

Hot Topic

You can't get colder than absolute zero. But there's no such limit to how hot something can be. Temperature measures the energy of movement of particles in a substance – and that energy can be incredibly high.

The official record for the highest air temperature on Earth is 56.7 °C, measured on July 10, 1913, at the appropriately named Furnace Creek Ranch in Death Valley, California. Furnace Creek also tops the charts for the highest ground temperature – a near-boiling 94 °C on July 15, 1972.

Only two other planets in the Solar System are hotter than Earth. Mercury, nearest the Sun, spins on its axis exactly three times for every two trips around its orbit – three days in every two years. During a Mercurian day, which, from sunrise to sunset, lasts 176 Earth days, the temperature rises to about 427 °C at the equator. At night, the lack of an atmosphere means that the temperature quickly plunges to about –180 °C. Venus, the second planet, is about twice as far from the Sun as Mercury, but manages to be hotter. Its dense carbon dioxide atmosphere acts like a giant greenhouse, maintaining a stifling surface temperature, day and night, of at least 462 °C.

Thousands of planets have been discovered around other stars and some of these have very small orbits and extraordinarily hot surfaces. About 1,400 light-years away lies the star WASP-12, similar in nature but somewhat bigger and brighter than the Sun. Circling around it, at a distance of just 3.5 million kilometres, or 1/43rd the distance of Earth from the Sun, is a planet with roughly one and a half times the mass of Jupiter. WASP-12b's orbit is so small that the gravitational pull of its central star has stretched the planet into an egg-shape and is stripping away 189 thousand trillion tons of its atmosphere every year.[1] The very substance of the planet is also being slowly consumed as its orbit decays and it spirals in towards total destruction in just a few million years' time.

Needless to say, WASP-12b is desperately hot. The heating effect of its star combined with that of the tidal forces that distort its shape conspire to raise the surface temperature to more than 2,200 °C – hot enough to melt iron. Yet WASP-12b doesn't glow, lava-like, as you might expect. On the contrary, its black, carbon-rich surface reflects just 6 per cent of the light that falls on it, making it as dark as asphalt.

Hotter even than WASP-12b – in fact the hottest planet known – is a world baked by the fierce rays of a star more than twice the size, and several times the mass, of the Sun. KELT-9 lies 670 light-years from Earth and shines with a brilliant white light due to its surface temperature of 9,900 °C – more than 4,000 degrees higher than that of the Sun.[2] Picture a world barely 5 million kilometres away from such a star, ten times closer than Mercury is to the Sun. The daytime side of KELT-9b sizzles at around 4,300 °C, hot enough to melt all substances on Earth and boil most of them. So hot is the hemisphere of KELT-9b which permanently faces its star that there's iron in the atmosphere, vaporised from the surface. This is a planet that's hotter than some stars.

The coolest of stars are the smallest, lightest red dwarfs, the surfaces of which sometimes barely reach 1,800 °C. At the other extreme are so-called Wolf–Rayet stars, one group of which are near the end of their lives and soon to explode as supernovae, while another group are comparative youngsters on the stellar scene. Both groups of Wolf–Rayets, though, are massive, very bright and very hot. The hottest of them all is WR 102, which lies near the centre of the Milky Way galaxy and belongs to an extraordinarily rare sub-species of Wolf–Rayet known as WO types, only ten of which are known. WR 102 is the hottest star yet discovered, with a surface temperature of about 210,000 °C.[3]

Deep inside, stars are much hotter than in their upper layers. At the centre of the Sun, temperatures around 15 million °C allow nuclear fusion to take place: hydrogen nuclei smash together with such ferocity that they combine to form helium nuclei with the release of huge amounts of energy. It's these kinds of temperatures, and more, that are needed for fusion to become a practical means of generating energy here on Earth.

The first experiments in fusion energy were carried out in the 1950s using small-scale devices to explore how ionised gas – plasma – behaves at high temperatures. In progressing towards practical sustained fusion, however, physicists faced a massive challenge: how to contain a substance at millions of degrees long enough to extract useful amounts of energy from it. Various approaches have been tried, including tokamak reactors, in which the plasma is confined in a doughnut-shaped ring by a powerful magnetic field. But it's only been within the past few years that the dream of harnessing fusion energy has come within our grasp.

In 2021, China's Experimental Advanced Superconducting Tokamak (EAST) set a new record for superheated plasma, sustaining a temperature of 120 million °C for 101 seconds and a peak of 160 million °C for 20 seconds.[4] That's much hotter than the centre of the Sun, but it needs to be because the plasma density is nowhere near as great. The trick is also to maintain super-high temperatures, above 150 million °C, for lengthy periods of time, in order to generate useful amounts of electricity. In terms of duration, EAST broke another record in 2021 by sustaining a plasma at 70 million °C for over seventeen and a half minutes.

On August 8, 2021, researchers at the Lawrence Livermore National Laboratory's National Ignition Facility (NIF) in California reported that they'd achieved the long sought-after state of 'ignition'.[5] For the first time, humans had made an artificial sun that was self-sustaining – in other words, that produced more energy than it took in. In 2022, the Joint European Torus (JET), near Oxford, broke new ground for the amount of energy generated by controlled fusion – 59 mega-joules in a five-second burst, equivalent to a power output of 11 megawatts.[6] If nuclear fusion can be tamed, it promises to deliver abundant safe energy without churning out greenhouse gases or leaving behind dangerous radioactive waste.

Inside the reaction chamber of the Joint European Torus (JET).

Temperatures far greater than those in fusion reactors have been created on Earth, but only for fleeting instants of time. In particle accelerators, physicists crash together some of the smallest constituents of matter at speeds approaching the speed of light. Momentary collisions at such high energies result in temperatures not seen in the universe since the time of the Big Bang. In fact, some of the experiments conducted in high-energy colliders are attempts to mimic the extreme conditions that prevailed in the first few instants of the cosmos.

In 2010, researchers at the Brookhaven National Laboratory on Long Island, New York, announced that they'd performed the hottest experiment on the planet. The aim of the PHENIX collaboration at Brookhaven was to recreate the so-called 'quark–gluon plasma' (QGP) that, it's believed, existed for several ten-millionths of a second after the universe's birth. By hurling gold atoms together at near light-speed, the scientists cooked up a QGP with a temperature of 4 *trillion* °C. Even this astounding record didn't last long, however.[7]

On the Swiss–French border, near Geneva, is the largest particle accelerator ever built – the Large Hadron Collider. Attached to it is ALICE (A Large Ion Collider Experiment), which fires lead ions at each other at 99.9999 per cent of the speed of light to form its version of a quark–gluon plasma. At the 2012 Quark Matter conference, news broke that ALICE had surpassed even PHENIX's temperatures by reaching 5.5 trillion °C.[8]

Experiments that mimic the Big Bang are all very well. But you can't beat the real thing when it comes to the hottest of the hot. The highest temperature that has any meaning, given our current understanding of physics, is the Planck temperature. This was reached a mere 10 million trillion trillion trillionth of a second after the universe began when the wavelength of thermal radiation was the shortest it could possibly be according to known science. For an ultra-brief instant of time the temperature soared to the highest it has ever been: a staggering 140 million trillion trillion °C.

CHAPTER 8

Sphere

PERFECT CIRCLES EXIST – but only in mathematics. Perfect spheres exist – but only in that abstract realm of thought where every point on a surface is *exactly* equidistant from the centre. The real world of matter is lumpy: it's made of lots of little bits, such as atoms and subatomic particles, rather than being indefinitely smooth. Perfection may be out of reach, but extraordinarily round objects do exist, both in nature and the world of artefacts.

People have been fashioning roughly spherical objects for thousands of years. Among the oldest known are stone spheres of different sizes found at locations around the Aegean and Mediterranean. Archaeologists aren't sure of their exact purpose but have theorised that they may have served variously as sling stones, throwing balls, or, in the case of small ones, counters for recordkeeping. Hundreds of rounded stones, not much bigger than marbles, were found at the Bronze Age town of Akrotiri on the island of Santorini. They're anywhere from 3,600 to 4,500 years old and, it's been suggested, could be pieces from one of the oldest board games ever created.

In the past decade or so, scientists have made artificial spheres of extraordinary precision. One motivation for achieving these examples of near-perfect roundness was to redefine the kilogram, the fundamental unit of mass in the International System of Units. For many years, the kilogram was defined as exactly equal to the mass of a small polished cylinder, made of platinum and iridium, which had been cast in 1879. The so-called International Prototype of the Kilogram (IPK) – aka *Le Grand K* – was housed under three nested glass cloches in a locked and climate-controlled vault in a Parisian suburb. Scores of near-identical copies of the IPK resided in countries around the world and served as reference masses for the accuracy of every measurement of mass or weight, whether in metric tons and milligrams or pounds and ounces.

But then scientists decided that some new and more reliable standard was needed. One of the options put forward was to make the roundest object ever fashioned by human hand. The challenge was taken up by the Australian Centre for Precision Optics and resulted in a ball, less than 10 centimetres in diameter and worth more than a million euros at the time, made from a single crystal of silicon-28 atoms.[1] Its surface is so free from imperfections that if Earth were as smooth, then the difference in height between the tallest mountain and the deepest ocean trench would be less than 5 metres.

Only one group of manufactured spheres are more perfect in shape than the silicon-28 orb. They are the ones constructed to fly aboard NASA's Gravity B spacecraft, launched in 2004 to test Einstein's general theory of relativity with exquisite accuracy. Four spheres, each 3.8 centimetres across and fashioned of fused quartz, served as gyroscopic rotors on this mission.[2] Their departure from perfect sphericity is less than two ten-millionths of their diameter, or the width of a few molecules. Scaled up to Earth-size the maximum difference in height of topographic features would be just 1.5 metres.

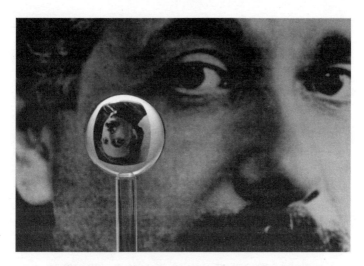

At the time, the fused quartz gyroscopes for Gravity
Probe B were the most nearly perfect artificial spheres
ever made. They differ from a perfect sphere by no
more than 40 atoms of thickness. One is shown here
refracting an image of Albert Einstein.

To find anything more flawlessly round than this we have
to leave Earth far behind. But before we head for the extreme
depths of interstellar space in search of the most perfect
natural sphere of all, let's take a quick look around our own
Solar System. Earth looks pretty round in photos but actually
bulges a bit around its equator. Jupiter and Saturn are notice-
ably squashed, for two reasons: they're mostly made of gas
and they spin around fast. Jupiter is eleven times wider than
Earth but rotates on its axis in less than ten hours. Saturn
spins slightly slower but, being less dense, is more flattened
because its rotational acceleration effectively cancels a larger
fraction of the planet's gravity at the equator. Its polar diameter is only 90 per cent of its equatorial diameter.

The Sun is also a gassy ball so we might expect to see some bulging of its equator as it spins around. Yet, remarkably, as measurements in 2011 by NASA's Solar Dynamics Observatory revealed, the equatorial diameter of the Sun is a miserly 0.0003 per cent bigger than its polar diameter.[3] Even granted the fact that the Sun rotates slowly, about once a month, scientists were surprised to find that it's a 99.9997 per cent perfect sphere. Shrunk to the size of a basketball, the Sun would be wider across its equator than pole-to-pole by just the width of a human hair. It remains an unsolved problem as to why the Sun is so very nearly perfectly round.

One thing is certain. Among all the countless trillions of stars in the universe the Sun isn't unique or even unusual. There must be many examples of stars that are as round or even rounder than the Sun. The difficulty is in proving this when the objects in question are so far away they appear only as points of light in even our largest telescopes. Fortunately, it isn't always necessary to see the shape of an object in order to be able to figure it out.

Case in point: KIC 11145123, a white star that lies about 3,900 light-years from Earth. It's unusual in that, although twice as big as the Sun, it spins three times more slowly. In 2021, a team of astronomers, led by Laurent Gizon from the Max Planck Institute for Solar System Research and the University of Göttingen, examined KIC 11145123 using a technique called asteroseismology. Stars display gentle oscillations, like the vibrations in air that give rise to sound or the standing waves of a plucked string. These oscillations can be studied to learn more about stellar interiors. They also reveal how much a star deviates from a perfect sphere. KIC 11145123, it turns out, is rounder even than the Sun – in fact, the roundest natural object found to date.[4] The star as a whole measures about 1.5 million kilometres across but the difference between its equatorial and polar radii is, incredibly, only 3 kilometres.

The ultimate in natural spheres still await discovery but a strong candidate is the smallest, most condensed form of star: a neutron star. The extreme gravitational pull of a neutron star has the potential to create the most nearly perfect shape if it were not for the fact that many of these objects spin extremely fast. The neutron star at the heart of the Crab Nebula, for instance, spins dizzily on its axis more than thirty times a second.

But as a neutron star ages and loses energy, its rate of rotation declines. This, in principle, should allow its self-gravity to pull its contents into a more and more nearly perfect spherical shape. In 2020, astronomers announced the discovery of PSR J0901-4046, a 5.3-million-year-old neutron star that takes seventy-six seconds – three times longer than its nearest rival – to complete one spin. Its properties, including its radio emission, which, puzzlingly, is detectable for only 0.5 per cent of each rotation, challenge our understanding of how neutron stars evolve.

Another contender for 'most perfect natural sphere' goes by the name 1E 161348-5055. It sits at the heart of a supernova remnant, 10,000 light-years away, and is by far the most slowly rotating neutron star known, with a period of 6.7 hours. Given that it's only about 2,000 years old, it's a mystery how it can be spinning so slowly.[5] However, if 1E 161348-5055 is what it appears to be – a ball of neutrons roughly 20 kilometres across, spinning at such a leisurely rate – it may be as perfect a sphere as the universe has to offer. The difference between its equatorial and polar diameters may be as small as the width of a single proton.

CHAPTER 9

Beyond the Superball

IN THE 1997 movie *Flubber*, Professor Philip Brainard, played by Robin Williams, creates a green goo that's spectacularly bouncy and elastic. Brainard's robot assistant describes the substance as 'flying rubber', which Brainard shortens to 'flubber'. One of its amazing properties is to increase speed every time it hits something, making it difficult to control.

In the real world, flubber, or anything like it, is simply impossible. No substance can gain energy from a rebound. Otherwise, you'd be able to use it to make a perpetual motion machine or generate unlimited amounts of free energy, and those are extravagances that physics won't tolerate.

A key factor in what happens when one object bounces off another is a quantity called the coefficient of restitution, denoted by e. It's defined as the ratio of the final to initial relative velocities before and after a collision. The maths of it were figured out in the seventeenth century by Isaac Newton and the result is often known as Newton's experimental law.

In the case of flubber, e would be bigger than one. But since we live in the actual universe and not some Hollywood science fiction caper, the maximum possible value for e is one, assuming a perfectly elastic collision. In fact, in all practical situations, some kinetic energy – energy of motion – is inevitably lost when one thing bumps into another and so the result is an *inelastic* collision, for which *e* is less than one.

Think about bounciness and the first thing that springs to mind is a ball. The whole idea of a ball, whether for play or sport, is that it rebounds in a fairly predictable way when it strikes, or is struck by, a surface. Not all balls are bouncy but those that are consist of some elastic material so that they maintain much of their momentum and kinetic energy after a rebound.

The first use of natural rubber has been traced to the Olmec culture of what is present-day Mexico, more than three and a half thousand years ago. This rubber, made from latex extracted from the *Hevea* tree, was used in making balls for the Mesoamerican ballgame, or *ōllamaliztli* as the Aztecs called it, a sport with ritual associations that's been played since at least 1650 BCE.

In the 1930s, Spalding, an American sports equipment manufacturer, originally based in Chicago, began producing its Hi-Bounce Ball, similar to a tennis ball but without the fuzzy felt covering and pink in colour. If dropped from a person's eye level, it would rebound to about half their total height. Nicknamed the Spaldeen or Pensie Pinkie, its convenient size, lightness and bounciness made it wildly popular in children's urban street games, such as Stickball, a street version of baseball.

In the mid-1960s, another recreational revolution began with the appearance of the first Superball. I remember being fascinated, as a slightly nerdy eleven- or twelve-year-old, by the incredible springiness of these things, especially when you threw them hard at the ground or a wall. It was as if something more than mere science was at work. In fact, the secret of the Superball was a new type of synthetic rubber invented by Norman Stingley, a chemist working for Bettis Rubber

Company in Whittier, California.[1] Experimenting during downtime on his own, Stingley mixed the synthetic polymer polybutadiene with some hydrated silica, zinc oxide, stearic acid and other ingredients. He then squeezed the resulting goo at a pressure of 3,500 pounds per square inch and vulcanised the compound with sulphur at 165 °C.

Stingley recognised straight away that he was on to something and that the exceptional bounciness of his new rubber had recreational potential. So, he took his invention to Dick Knerr and Arthur 'Spud' Melin, owners of Wham-O Inc., who had a reputation for creating innovative toys based on quirky, often-accidental inventions. Sure enough, Knerr and Melin saw a future in Stingley's surprising substance and requested that he go back to his lab to make some final tweaks. The finished material, which Stingley called Zectron, was black and manufactured into pocket-sized spheres that were just under 5 centimetres in diameter. The Superball, 'made from amazing Zectron', as the adverts announced, was a runaway success, racking up 20 million sales for Wham-O between 1965 and 1970.

Wham-O claimed the Superball would bounce six times higher than a tennis ball if dropped from the same height. Thrown with enough force at the ground it could bounce over a three-storey building – a lot of return on a 98-cent investment. A Superball lost only 10 per cent in height after each rebound, giving it an impressive e-value of about 0.9.

Bouncier still – at least according to its manufacturer, Maui Toys – is the Skyball. Much larger than a Superball, with a diameter of 10 centimetres and with a hollow interior containing a mixture of helium and compressed air, it's advertised as being able to bounce 23 metres straight up.

We tend to think of bouncy materials as being stretchy and squishy: if not actually rubber then at least rubberlike in their consistency. But there are substances that are surprisingly springy yet extremely rigid. A glass ball, for instance, will bounce higher than an ordinary rubber ball of the same size, providing it doesn't break. A steel ball, such as a ball bearing, dropped on a hard floor, also rebounds more than a rubber one.

There are obvious safety reasons why we don't use glass or metal projectiles in ball games. But in terms of bounciness they can sometimes win out because of the physics involved. If you drop a ball, it bounces well only if it gets back most of the energy that it temporarily imparts to the surface. In the case of a rubber ball, it compresses and deforms before returning to its original shape, which uses quite a lot of energy. A rigid ball, like one made of glass or metal, hardly changes shape at all so that most of the energy of the impact is retained in the rebound. Of course, how the impacted surface behaves is also a major factor. A ball bearing, for instance, doesn't bounce well from a soft surface such as grass because almost all its kinetic energy is lost in deforming the surface.

One of the toughest yet springiest substances yet encountered by scientists goes by the undistinguished name SAM2X5-630. It's a strange, glass-like material, stronger than steel but internally composed of atoms in a jumbled state, the way they are in glass.[2] SAM2X5-630 belongs to a family of substances called bulk metallic glasses whose extraordinary properties are only just beginning to be exploited in manufacturing. A phone made of such a material would be tough enough to survive a fall from a three-storey building and be bouncy enough to return almost to your hand. The contact surface of golf clubs made of bulk metallic glass can be used to strike balls further, while a spacecraft whose outer surface was coated in such a material could easily deflect any impacting debris.

CHAPTER 10

Tesla Max

MAGNETS CAN BE surprisingly powerful. Hold a child's magnet over a paperclip resting on a table and the paperclip will jump up and stick to the magnet. That means the upward force of the little magnet acting on the paperclip is stronger than the entire downward force of Earth's gravity trying to keep the clip on the table.

Earth has its own magnetic field. Generated by movements within our planet's liquid nickel-iron core, it's what deflects compass needles and serves as a protective shield against high-energy charged particles coming from the Sun. But Earth's magnetic field at the surface is quite weak. Even a small handheld magnet is much stronger at close range and enough to make a compass needle spin around.

The strength of a magnetic field is measured in teslas (T), named after the Serbian American electrical engineer Nikola Tesla. In fact, this is a very big unit, and the strength of most ordinary magnets, natural or artificial, is measured in small fractions of a tesla. For example, Earth's magnetic field strength at the surface varies from about 30 microtesla (millionths of a tesla) at the equator to about 65 microtesla at the north and south geomagnetic poles. That's similar to the strength of the magnetic field you'd pass through if you walked under a high-voltage power line.

Magnetism was discovered in the first place, thousands of years ago, because of naturally occurring magnetic minerals. The commonest and best known of these is lodestone, which is a magnetised magnetite, a form of iron oxide. A number of different minerals, mostly containing iron, are attracted by magnets, but only a few, such as lodestone, are found to be magnetised themselves.

Permanent magnets, like those you can stick on a fridge door, create a persistent magnetic field. They come in various shapes, including bars, rings and horseshoes, and are found in all kinds of devices from headphones to cars. The most familiar kind are made of ferrite, a ceramic material produced by mixing iron oxide with small amounts of other metals like nickel and manganese. A typical fridge magnet has a strength of about five milliteslas, or five thousandths of a tesla.

The strongest permanent magnets are made from an alloy of the rare-earth element neodymium with iron and boron.[1] Developed independently and at about the same time in 1984 by General Motors and Sumitomo Special Metals, neodymium magnets can produce a magnetic field of up to 1.25 T near their surface. That great strength has led to them replacing other types of permanent magnet in a huge number of applications from computer hard drives to mobile phone speakers and cordless power tools. They've also inspired new uses, including magnetic building sets for children. But their immense strength at close range can be a health hazard.

Neodymium magnets bigger than a few cubic centimetres are powerful enough to crush fleshy parts of the body or break bones that become pinched between two of them. Cases have occurred where young children have swallowed several magnets and been injured when sections of the digestive tract were squeezed shut by them.

In 1820, the Danish scientist Hans Christian Oersted discovered that electric currents create magnetic fields. Four years later the English physicist William Sturgeon invented the electromagnet. His first effort was a varnished, horseshoe-shaped piece of iron wrapped with copper wire. When a current was passed through the copper coil, the iron, insulated from the wire by its layer of varnish, became magnetised. When the current was stopped, it instantly lost its magnetisation. Sturgeon's original electromagnet was relatively feeble but it could still lift twenty times its own weight.

Today, electromagnets of various kinds are the most powerful magnets available and have numerous applications. They're an essential part of medical magnetic resonance imaging (MRI) systems, in which they generate fields of up to 3 T. MRI scans, unlike X-rays, involve no exposure to radiation so are completely safe in this regard. The only risk they pose is if the patient has anything containing iron, or other metals that can be magnetised, in their body. You certainly wouldn't want to enter an MRI machine if you had a pacemaker, cochlear implant, or any other iron-based implant.

Accidents with MRI machines are extremely rare. Usually, the worst that can happen is a burning sensation, for example if a patient has tattoos which sometimes have iron oxides in their ink. Fatal incidents are almost unheard of. One took place in 2018 in Mumbai, India, when a man was sucked into an MRI machine while visiting a relative in hospital. He'd been asked to carry a metal oxygen cylinder by a junior staff member who assured him the machine was switched off when, in fact, it wasn't. The victim was yanked into the side of the device, which ruptured the tank and led to his death from inhalation of the released liquid oxygen contents.

A little-known fact is that almost everything – not just iron and a few other metallic and rare-earth elements – gives rise to magnetism. This isn't the strong type of magnetism, called ferromagnetism, with which we're familiar. The very weak field produced by most things, including our bodies, is known as diamagnetism. Its existence means that, given a strong enough magnet, there can be an interaction with an object's feeble diamagnetic field. In 1997, scientists used this principle to levitate a frog.[2] The unsuspecting amphibian was placed in a super-strong magnetic field of 16 T and made to rise a few centimetres in the air. Though no doubt surprised by its temporary ability to defy gravity the animal seemed to suffer no lasting effect from its aerial adventure.

All the strongest artificial magnetic fields today, including the one in the frog levitation experiment, are produced by electromagnets that use superconductors. These are materials through which an electric current can pass with zero resistance. Superconducting magnets are made by winding wire made from a material such as niobium–titanium or niobium–tin into a coil. The coil is then cooled to very low temperatures using liquid helium or liquid nitrogen so that the material becomes superconducting. When a current is passed through the coil, it creates a powerful magnetic field.

Among their many applications, superconducting electromagnets bend the paths of charged particles in high-energy accelerators. In a test, a demonstrator magnet, designed and built by a team at the Department of Energy's Fermilab, near Chicago, achieved a 14.5-tesla field strength for an accelerator steering magnet. In lab experiments even more powerful fields have been produced.[3]

In 1999, scientists at the US National High Magnetic Field Laboratory (MagLab) in Tallahassee, Florida, ran intense electric currents through coils made of a cuprate semiconductor in which layers of copper oxide alternate with layers of other metal oxides.[4] Their equipment, known as a hybrid magnet, involved a second component called a Bitter magnet made from circular conducting metal plates and insulating spacers stacked in a helical arrangement. These two ways of producing a magnetic field each have their disadvantages. Superconducting magnets need little power but have an upper limit on magnetic field strength. Bitter magnets, on the other hand, consume a lot of power but can produce really strong fields. In combination they cancel each other's weaknesses to some extent, allowing steady, super-strong fields to be generated. The MagLab team produced a field of 45 T – a record that was pipped in 2022 by physicists at the Steady High Magnetic Field Facility in Hefei, China, who, with their hybrid system, reached 45.22 T.[5] That's nearly one and a half million times stronger than Earth's magnetic field at the equator. Pulsed magnets can generate stronger fields but can sustain them for only a fraction of a second.

There's no known limit, in theory, to how strong magnetic fields can be. But human-made magnets do face a barrier. An electric charge travelling along a magnetic field line moves in a spiral path and is pushed away from regions where magnetic field lines are bunched together. The stronger the magnetic field, the faster and tighter a charge spirals around, and the harder it gets pushed away from regions of high magnetic field gradient. Everything around us on Earth is made of atoms, which themselves contain charged particles – protons and electrons. So, if magnetic fields are strong enough they can deform and even destroy ordinary objects. The limit is about 50 T: a machine that could create a continuous magnetic field stronger than this would rip anything – including itself – to pieces.

But objects that aren't made of ordinary matter face no such limitation. At the end of its life, when a massive star explodes, it may leave behind an incredibly dense object called a neutron star. All neutron stars have strong magnetic fields of at least 10,000 tesla, but some, known as magnetars, have what is believed to be the most powerful magnetic fields in the universe. The magnetar phase lasts only 10,000 years or so but during that time the field strength at the surface of the fantastically dense object may be up to 100 million tesla – about a trillion times more powerful than the field surrounding Earth.

CHAPTER 11

Stop Right There

THE CUBE, ABOUT 10 centimetres wide, is shining bright with an orange-yellow light. It's just been taken out of a furnace at 1,200 °C – a temperature hot enough to melt copper or gold. Yet now, just seconds later, a demonstrator is lifting up the glowing shape with her bare fingers and showing it to the startled audience. How is it possible to hold an object that's much hotter than the hottest element on an electric oven without being burned?

The cube in this demonstration is composed of LI-900 – the same stuff used to make the tiles that covered much of the Space Shuttle and that protected it from the heat of re-entry into Earth's atmosphere.[1] Most of the Shuttle's airframe was made of aluminium, which is strong and lightweight but begins to soften at temperatures as low as 175 °C. The layer of tiles ensured that hardly any of the heat generated by friction with the atmosphere penetrated through to the airframe to weaken it.

LI-900 is an incredibly good thermal insulator. Put another way, it has a very low thermal conductivity: it transmits hardly any heat, which is why it can be handled even when it's been exposed to very high temperatures. It consists almost entirely of silica fibres and air. Silica fibres are long slender threads of sodium silicate. In the Shuttle tiles, just 6 per cent of the volume is occupied by fibres and the rest is air to keep the weight to a minimum.

Thermal conductivity is measured in units of watts per metre-kelvin (W/m.K) – the lower the number, the less heat is allowed to pass through the substance. Metals are among the best conductors of heat. Gold comes in at 315 W/m.K and silver at 429 W/m.K. Both, however, are easily surpassed by diamond which, as well as being the hardest material on the planet, also has the highest thermal conductivity of any substance: 2,000–2,500 W/m.K, depending on the purity of the crystal.

At the other end of the scale, the thermal conductivity of air is just 0.024 W/m.K. That's why some of the best insulating materials have large spaces in them filled with air. To prevent heat escaping from our homes we put in loft insulation often consisting of rolls of fibreglass, mineral wool, or natural fibres such as sheep's wool. Some of the best thermal insulators are those that have evolved over many millions of years to keep animals warm: the fur of mammals and the feathers of birds, both of which trap pockets of air. But the best thermal insulator of all is a human-made substance.

As its name suggests, aerogel is a combination of a gel and air or some other gas. Normally, gels are mostly liquid but behave like solids because of the three-dimensional crosslinking of their molecules. In aerogel the gel is replaced with a gas while leaving the gel structure intact. The commonest type, silica aerogel, has an extraordinarily low thermal conductivity, from about 0.03 W/m.K at atmospheric pressure to as little as 0.004 W/m.K in a vacuum. It's also translucent and amazingly light, earning it the name 'frozen smoke', yet surprisingly strong – able to support objects thousands of times its own weight.

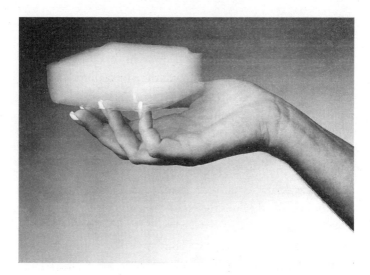

A block of silica aerogel.

Often, substances that insulate well in one regard also insulate well in another. Glass, for instance, is a good thermal insulator and a good electrical insulator. This is true of most materials: either they conduct both heat and electricity well or they insulate well against both. The common factor is the availability of free electrons – electrons that aren't bonded to atoms or molecules and so can move about freely. Electricity is due to a current of (negatively charged) electrons. But what is often not appreciated is that the movement of electrons also plays an important role in the conduction of heat through a solid such as a metal.

Electrical conductivity is measured in something called siemens per metre (S/m) – the higher the number, the better a substance is at allowing electricity to flow. Silver has the highest electrical conductivity of all with a score of 6.3×10^7 S/m, closely followed by copper with a conductivity of

5.96×10^7 S/m. Since copper is so much cheaper than silver this explains why it's used for most electrical wiring.

As everyone knows, water and electricity make a dangerous pair. A swimming pool isn't the place to be if there's a storm about and damp electrical sockets are best avoided. Pure water, though, is actually a very good electrical insulator. What makes most of the water around us, from drinking water to seawater, good at conducting isn't the water itself but the various salts and other substances that are dissolved in it. Seawater is the most conductive of all because it's rich in sodium chloride and other salts. But, oddly enough, you're less likely to be electrocuted if there's a lightning strike in the sea than in a swimming pool or a bath. The reason for this is that seawater is such an efficient conductor of electricity that an electric current, which always chooses the line of least resistance, may effectively ignore a human body in the water and stick to a more efficient pathway – namely, through the charged particles of dissolved salts.

Air is an excellent electrical insulator and will only conduct if forced to do so by a high voltage. Several thousand volts are needed to make a spark jump across a gap of just a millimetre or so. A typical lightning strike creates a conducting path from cloud level to the ground, but to break air down over that kind of distance takes on the order of a gargantuan 300 million volts.

Plastics, rubber and wood are other examples of materials that insulate well against both heat and electricity. But there's one glaring exception to the link between thermal and electrical conductivity. As we've seen, diamond is the best thermal conductor of any natural substance, yet it's not a very good electrical insulator. To find an explanation we have to descend to the molecular level. Diamond consists of carbon atoms linked together by strong bonds that involve shared electrons. This rigid molecular framework is very effective at transmitting vibrations that carry thermal energy. But the absence of any free electrons means that electricity can't flow through. By contrast, another form of carbon, graphite, does have free electrons and so combines good thermal and electrical conductivity.

Sound is another phenomenon that can pass with greater or lesser ease through different materials. As a general rule, sounds are conducted better by denser substances, such as metals and concrete. Most of the sound we hear travels through air to reach our ears, so we think of air as being a normal – and therefore effective – carrier of sound waves. But, in fact, air is a poor sound conductor as are a lot of materials that contain numerous pockets of air, such as foam, mineral wool, cardboard, feathers and cork. Good thermal insulators often make good acoustic insulators as well because of their low density.

Sound must have a medium of one kind or another through which to travel because it's due to the back and forth oscillation of atoms and molecules. In the absence of matter sound is stopped in its tracks: a vacuum is a perfect acoustic insulator.

CHAPTER 12

The Persistence of Sound

ABOUT A DOZEN miles from where I live nestles the town of Cupar in Fife, Scotland. The tallest structure there is a 60-metre (197-foot) high concrete silo, originally built in 1964 as a storage facility for the adjoining sugar beet factory. Now empty, the Cupar Silo was the site of an experiment in 2014 which sought to explore an outstanding property of the building: the cavernous interior has an extraordinarily long reverberation time of 36.5 seconds.[1]

Reverberation time is how long it takes sound in an indoor space to die away after the source of the sound has stopped. To make it a standard measure, it's defined as the time taken for a sound to fade by 60 decibels below its original level. Imagine clapping your hands and still being able to hear faint traces of the noise more than half a minute later. The extreme acoustics of the Cupar Silo are an accidental artefact of its internal dimensions, shape and reflective concrete walls. But it isn't unique in being highly reverberant. Caves, cathedrals and abandoned factories are among other spaces where sound can bounce back and forth many times before it finally dies away. The record reverberation time for any structure on Earth was also recorded in Scotland.

In 1941, work was completed on a series of huge tanks deep underground near Invergordon, Ross-shire. Known as the Inchindown oil tanks, they were built as a wartime bombproof oil storage depot for the Royal Navy. Each of the five main tanks is 237 metres long – longer than two football fields – and 9 metres wide, with an arched roof 13.5 metres high. University of Salford acoustic engineer Professor Trevor Cox heard about the facility when it was featured on BBC's *The One Show* and realised it would be an interesting site for some reverberation tests. In 2014, he recorded the sound from a pistol blank fired inside one of the tanks. The last sound faded out 112 seconds after the trigger was pulled, making it the longest lasting reverberation ever recorded.[2]

When playing music indoors a little reverberation is a good thing but too much can lead to a confusing cacophony. In some large Gothic cathedrals it can take up to nine seconds for the sounds from a pipe organ to die away, so that several notes, although played separately, can end stacked on top of one another in a discordant pile-up. Concert halls are designed to avoid this problem, although performers recognise differences between reverberant spaces, describing some as being very 'live' and others as 'dead'. Optimum reverberation times depend on the type of music for which the hall is intended. Generally, good concert halls have a reverberation time between 1.8 and 2.2 seconds at mid-frequencies.

One of the features of the human ear is that it shows progressively greater discrimination against low frequencies as the sound gets softer. As a result, listeners far away from a source – say, an orchestra – may conclude that bass notes are too quiet in the overall mix. One way to offset this bias is to design auditoriums in such a way that the reverberation time for low frequencies is greater than for high frequencies. This gives a natural-sounding boost to bass sounds for audience members at the rear of the space. In practice, a longer reverberation time for low sounds often happens naturally, particularly if a lot of wood is used in the construction since wood absorbs higher frequencies more than lower ones.

The Royal Albert Hall in London is famed for its architectural grandeur and notorious for its dodgy acoustics. Completed in 1871, this magnificent domed building has hosted everything from large classical concerts to modern pop and rock gigs. But it was designed by mid-Victorian royal engineers, not by architects and acousticians with an understanding of twenty-first-century science. It quickly became obvious that sounds bounced around its circular walls and, especially, its great dome, far too much. The Albert Hall was both overly reverberant and echoey.

Reverberation and echoes are related but aren't the same. Reverberation is the result of lots of reflected waves, which, when they're processed by the brain, are interpreted as a continuous, complicated sound. An echo, on the other hand, is a distinct but quieter copy of the original caused when a pulse of sound comes back to the listener after bouncing off a surface. If there's a delay of more than 50 milliseconds between the first and the second sounds reaching the ear, then they'll be perceived by the brain as separate rather than as one extended event. In terms of distance, an echo can be heard only if the reflecting surface is at least 17 metres away.

From the very first concert, following the opening ceremony, it was clear that the Albert Hall had a problem. Its concave, glass dome reflected and focused any sound that fell upon it, creating an obvious echo. A joke began to circulate that the venue was the only place where a composer could be sure of hearing their work played at least twice. In the century and a half since, numerous efforts have been made to improve the acoustical performance. Most recently, between

2017 and 2019, more than £2 million was spent on a major renovation, including a computer-designed sound system that can adjust the level of reverberation and echo perceived by different sections of the audience. Fifteen kilometres of cable were needed to connect the many new speakers that were installed. While laying cables below the original flooring, some interesting artefacts were unearthed – among them, a nineteenth-century beer bottle probably discarded by a worker during the building of the Hall.[3]

Another famous domed London landmark has its own acoustic peculiarity. I'm one of many people who've visited the Whispering Gallery, a circular walkway at the base of the dome 30 metres above the crossing of the nave of St Paul's Cathedral. The architect, Sir Christopher Wren, hadn't intended to create a sonic talking point – literally or metaphorically – when he designed the structure. But the gallery quickly became a fashionable meeting place after the consecration of St Paul's in 1708 with the discovery that by whispering along the curved wall your words can be heard by someone with their ear to any other part of the wall, even if they're on the opposite side of the walkway, more than 33 metres away.[4]

The correct explanation of the phenomenon was provided by physicist Lord Rayleigh in 1878. 'Whispering-gallery waves', as they've come to be known, Rayleigh realised, are best generated by someone whispering *along* the curving wall rather than into it. The sound then travels along a series of straight-line paths or chords (lines joining different points on the circumference) of the round gallery. By clinging to the walls in this way, the sound falls away in intensity only as one over the distance rather than as one over the square of the distance in the case of sound which spreads out in all directions. This gentler drop-off accounts for why whispers are audible all the way round the gallery.

The Whispering Gallery inside the dome of St Paul's Cathedral.

We find this phenomenon at work in buildings around the globe. The dome of Gol Gumbaz, a seventeenth-century mausoleum in Karnataka, India, has an exterior diameter of 44 metres and, as with St Paul's, an accessible walkway at the base of the dome where the whispering effect can be heard. The Imperial Vault of Heaven, a circular structure that's part of a complex of fifteenth-century religious buildings in the southeastern part of central Beijing, houses the Echo Wall which also transmits sounds around its circumference.

Echoes are common both inside buildings and in the natural world. Anywhere that a wall, made of rock or some hard artificial material, some distance away, provides a reflecting surface will create the effect. The Grand Canyon in Arizona and Echo Point Lookout, near Katoomba, in the Blue Mountains, Australia, are classic examples.

Not all the effects associated with reverberation and echoes have been properly explained. One of the strangest sound anomalies in the world can be heard at El Castillo, also known as the Temple of Kukulcán, in Chichen Itza, once a city and now an archaeological site, in Yucatán, Mexico. Built by the Mayans between the ninth and twelfth centuries, this Mesoamerican step pyramid is part of a larger complex that includes a ball court, temples and other buildings. To experience the weird acoustic effect, you simply stand in the field that faces the exterior steps of El Castillo and clap your hands once. An instant later an echo will come back that sounds, not like a clap but instead something much more like a bird chirp – and no one quite knows why.

CHAPTER 13

Gee Whiz

THE INNOCUOUSLY NAMED Flip Flap Railway at Paul Boyton's Sea Lion Park in Brooklyn, New York, was one of the most dangerous fairground rides ever built. Completed in 1895, it was the first looping rollercoaster in North America (earlier 'centrifugal railways', as they were called, having been constructed in Europe in the middle of the century). Flip Flap Railway featured a 25-foot-diameter vertical circular loop – and therein lay the problem: the loop was very small and perfectly round. It meant that as riders went into the circular section, and again as they came out of it, they were subjected momentarily to a bone-crushing 12 g acceleration, enough to cause serious neck and spine injuries.[1]

Nowadays, amusement ride designers are savvier and take both physics and human physiology into account in their more intense engineered-for-thrills schemes. Loop-the-loop sections of coasters are made not circular but teardrop-shaped, following a curve that mathematicians call a clothoid, so that riders can enter and exit the inverted section at more moderate accelerations.

G-force is a measure of acceleration that produces a feeling of increased weight. So, if you're undergoing an upward acceleration of 2 g, you'll feel twice as heavy as normal. Or, if you're in a car that's accelerating at 1 g, you'll be pressed back in your seat with a force that's equivalent to your normal weight. The only time most of us will experience more than a couple of g's is if we ride aboard a rollercoaster or other high-speed theme park attraction.

At the time it opened, in 1978, Shock Wave, at Six Flags Over Texas, near Arlington, offered its riders the maximum acceleration of any rollercoaster on the planet. After an initial 35-metre lift, followed by a 180-degree right-hand bend and a slight dip, the train goes down the first drop into back-to-back loops with peak forces of 5.9 g. It was pushed into second place in the g-force league table when Tower of Terror at Gold Reef City, Johannesburg, South Africa, went into operation in 2001. This white-knuckler lifts each 8-passenger car up with an elevator lift then pushes the car forward into a vertical drop which dives 15 metres underground into a mine shaft and subjects its riders to a maximum of 6.3 g.

The most extreme – and purely hypothetical – rollercoaster ever conceived is the Euthanasia Coaster, designed in 2010 by Julijonas Urbonas, a Lithuanian artist and PhD candidate at the Royal College of Art in London. This ultimate ride would intentionally kill all of its passengers or, as Urbonas put it, take lives 'with elegance and euphoria'. Those travelling on the Euthanasia Coaster would climb to a height of 500 metres before plunging at speeds of up to 360 kilometres per hour into a series of seven successive loops of decreasing radius crafted to ensure a full minute's worth of exposure to a least 10 g. During that time, the brains of those aboard would be deprived of blood (and therefore oxygen) leading to unconsciousness and, finally, death.

On an entirely more serious note, GLOC – g-induced loss of consciousness – became a known danger during World War II for pilots performing high-speed manoeuvres. Some way was needed to test the effectiveness of anti-g suits designed to prevent aviators from passing out when subjected to high levels of acceleration. This led Canadian scientists Wilbur Franks and Frederick Banting to set up the first reasonably powerful human centrifuge at the Canadian military's Clinical Investigation Unit (CIU) in Toronto in 1941. The project was supposed to be secret but it was clear that something unusual was going on in the building where the device was housed. The 200-horsepower motor that drove it shared a power line with the rest of the neighbourhood, and every time it was fired up it drained current from the overhead lines supplying streetcars on the road outside so that they ground to a halt.

The CIU centrifuge had a round gondola hung from a horizontal arm that was attached to a vertical shaft. The motor spun the shaft, causing the gondola to swing out so that it was tilted at almost 90 degrees when moving fast. The seat inside was like that of a fighter plane and suspended independently of the gondola, so that the rider could be positioned at different angles – even upside-down to produce negative g.

An observer in the control room sent signals to the gondola by turning on lights and sounding a buzzer; the rider replied by switching the signals off. If he turned off only the buzzer it meant he was still conscious but had suffered a blackout – a total loss of vision. If he failed to turn off the lights or the buzzer, it was the sign that he'd become unconscious. The results of the CIU centrifuge helped Franks develop the first operationally practical g-suit.

The much larger Johnsville centrifuge, at the Naval Air Development Center (NADC) in Warminster, Pennsylvania, played a key role in testing astronauts for the manned space programmes. But its first objective, after it was completed in July 1950, was to study the effects of g-forces produced by high-performance aircraft and, in particular, to help train pilots for flying the X-15 rocket plane.

Before the start of the Space Age, no one knew if the human body could stand up to the harsh treatment it would receive getting into and out of space, not to mention what the environment of space itself would do to a man or woman. In being blasted into orbit atop a powerful rocket, an astronaut would feel for minutes at a stretch much heavier than on Earth. During re-entry, the g-forces would be even higher. Through experiments on the ground and the experiences of pilots during extreme aircraft manoeuvres, it was known that people could withstand brief exposure to very high g's. But how much space travellers could take over longer periods, while accelerating into orbit and decelerating upon their return to Earth, remained a mystery.

The Johnsville centrifuge, known as 'The Wheel', was housed within a round, cavernous building at the NADC. The site was chosen because the turning forces the centrifuge could produce were so strong that the giant machine had to be anchored directly to bedrock to stop it shaking loose, and Warminster had some of the most stable bedrock engineers could find.

The steel gondola of the centrifuge was shaped like a flattened sphere, measuring 10 feet by 6 feet, and was mounted on a 50-foot arm, at the other end of which sat a 4,000-horsepower engine. So powerful was this motor that, flat out, it could whip the gondola to speeds of 175 mph in just seven seconds, reaching a potentially deadly maximum of 40 g. Equipped with dual gimbals, the gondola could be rotated so that the test subject was oriented in various positions relative to the applied g-force.

A lot of the time, the men who rode the Wheel were ordinary staff members at Johnsville or other Navy personnel who put themselves forward. The big names of aerospace were too busy flying to spend long hours being research guinea pigs, especially in the early days of the Wheel when its performance was being evaluated. Among the local volunteers was aerospace medical technician Art Guntner who, in his time at NADC, climbed into the belly of the beast some 350 times and briefly soaked up to 15 g of punishment – an acceleration that dwarfs the 4 to 6 g max experienced aboard top-fuel dragsters or modern human-rated spacecraft.

It was Guntner's job to prep the Mercury Seven – the original seven astronauts selected to take part in the Mercury space missions – before they rode in the centrifuge. He helped brief them on the results of the early simulations and his own experiences at high g. The gondola was kitted out with an instrument panel and hand controller like that in the Mercury capsule, and could be depressurised to the actual flight pressure of 5 pounds per square inch. Test subjects would report on how well they could use the controls while under high-g loading and describe any adverse effects they felt. An especially valuable but frightening feature of the gondola was that it could be put into a tumble in which the accelerations might lurch gut-wrenchingly from positive 10 g to negative 10 g in roughly a second.[2]

Under the crushing force of 6, 8 or even 10 g, normal breathing isn't an option. Inhaling in the usual way, when it feels as if you weigh half a ton or more, is out of the question. '[I]t's impossible to reinflate your lungs,' wrote Apollo 11 Command Module pilot Michael Collins, 'just as if steel bands were tightly encircling your chest. So you have to develop an entirely new method, keeping the lungs almost fully inflated at all times, and giving rapid little pants "off the top".'

John Stapp aboard the *Sonic Wind 1* rocket sled
slowing suddenly from a speed of 632 miles per hour.

Those given the green light to ride varied a lot in how well they tolerated the g-forces of the Wheel. Some people not only coped well with the centrifuge but strove to test its limits – and their own. One of these hardy souls was Navy Reserve officer Carter Collins who, in August 1958, withstood a crushing 20 g for six seconds using a grunting technique to avoid blacking out or suffering chest pains.

But even these feats were surpassed by those of US Air Force flight surgeon John Paul Stapp who, in the 1950s, subjected himself to astonishing rates of acceleration and deceleration aboard rocket-powered sleds.[3] His final and most extreme run took place on December 10, 1954. He was strapped, forward-facing, into an exposed chair on the back of a rocket-sled, called *Sonic Wind 1*, that could travel faster than a Jumbo Jet. Ahead of him was a 2,000-foot-long rail track stretching across the arid landscape of Muroc Army Air Field in California. Stapp himself counted down, his voice sounding over the intercom in the control room some distance away. At 'zero', the rockets were ignited and the sled accelerated ferociously over the next five seconds to an astonishing 632 miles per hour, demolishing the land-speed record.

Moments later, the sled ploughed into a trough of water, which brought it to a brutally abrupt halt. No human before or since has willingly undergone such a jolt: from over 600 mph to rest in a little over a second – the equivalent of hitting a brick wall at 120 mph. So fast had Stapp travelled that dust particles had speared through his flight suit, raising blisters on his body. So suddenly had he slowed down that the capillaries in his eyes burst, his eyeballs bulged from their sockets, and he was left temporarily blinded. Stapp had maxed out at an astonishing 40 g deceleration and, despite being battered and bruised, had survived to tell the tale.

SPACE

CHAPTER 14

Mega-worlds

MORE THAN 1,300 Earths would fit inside the largest planet in the Solar System, Jupiter. It's 318 times as massive as Earth and 2½ times as massive as all the other planets orbiting the Sun put together. A single storm on Jupiter, known as the Great Red Spot, has been raging for centuries and is wider than our own world. But going around other stars are planets even bigger than Jupiter. So, just how large can planets be?

Eight known planets orbit the Sun. The inner four are small, rocky worlds like Earth. Then come two gas giants, Jupiter and Saturn, and beyond them two ice giants, Uranus and Neptune. Pluto was, controversially, demoted to 'dwarf planet' status at an international meeting of astronomers in 2006.

The first extrasolar planets, or exoplanets, were discovered in 1992. But these turned out to be oddballs, circling around not ordinary stars but pulsars – dense stellar remnants of supernovae. It was 1995 before the first planet around a Sun-like star, 51 Pegasi, 50 light-years from Earth, was confirmed. 51 Pegasi b turned out to be an entirely new type of world. It was a 'hot Jupiter', in other words, a gas giant in an extremely small orbit around its central star. A high percentage of the early exoplanets to come to light fell into this same category, for the simple reason that the detection method most commonly used at the time favoured objects that caused the biggest wobbles in the motion of their parent stars – massive planets in tiny orbits.

Although Jupiter dominates the Sun's planetary system, astronomers realised there were almost certainly bigger and more massive planets elsewhere in space. This was confirmed in 1996 with the discovery of 47 Ursae Majoris b, which boasts about two and a half times the mass of Jupiter and was also the first known exoplanet not to have an unusually small orbit.

Theorists began to speculate more keenly on where the cut-off point for big, massive planets might lie. Jupiter had sometimes been described as a 'failed star' in that it was largely made of the same elements – hydrogen and helium – as a star like the Sun but lacked sufficient mass to ignite fusion reactions in its core. If, in the early stages of its formation, an object managed to pull in much more gas and dust from its surroundings, its self-gravity might squeeze and heat its central region enough to trigger nuclear fusion there. But at what point did a planet stop being a planet and become a star?

By 2011, researchers had discovered about 180 'super-Jupiters' – planets more massive than Jupiter – some hot, some cold, depending on their orbit. Much greater mass, however, doesn't necessarily imply proportionately greater size. The more massive a gas giant is, the more strongly it pulls itself together by gravity, so that its density may be much greater than that of Jupiter but not necessarily its diameter. A case in point is Upsilon Andromedae d, one of several planets found in orbit around a star that lies 129 light-years from the Sun. Estimates place its mass at around seven times that of Jupiter but its diameter only about 20 per cent greater.

As long ago as the 1960s, astronomers had theorised that there might exist objects intermediate between planets and stars, which later became known as brown dwarfs. These would have sufficient mass to squeeze their cores hard enough to initiate a kind of nuclear reaction known as deuterium fusion. This happens at lower temperatures than the normal hydrogen fusion responsible for generating the light and heat of stars like the Sun. Theory suggested that an object weighing more than about thirteen Jupiters could kickstart deuterium fusion and become a brown dwarf. If its mass exceeded about eighty times that of Jupiter, though, it would be able to fire up normal hydrogen fusion in its core and shine as a fully-fledged star.

Lying 133 light-years from Earth is a star called HR 8799. It's hotter, a bit bigger and somewhat more massive than the Sun, and is just visible to the naked eye in the constellation Pegasus. Four super-Jupiters have been found orbiting it, all at least five times as massive as Jupiter, and one that's about nine times as massive.[1] Despite their great weight and size, the mega-worlds of HR 8799 are unquestionably planets. In other cases, the status of an object, whether a giant planet or a brown dwarf, is uncertain.

Take, for instance, NGC 2423-3 b. It goes around a red giant, part of a cluster of stars about 2,400 light-years away. Astronomers found it because of wobbles – slight back-and-forth movements – of the star caused by its companion as it tugs first one way and then another in its orbital journey. We know that the mass of NGC 2423-3 is at least 10.6 times that of Jupiter. This lower estimate comes from assuming that the plane of the companion's orbit around the red giant lies directly along our line of sight. If the orbit is tilted away from our line of sight, the companion's mass would have to be greater to account for the observed wobbles in the parent star. Its mass could, in fact, be big enough to push it into brown dwarf territory.

Similar uncertainty hangs over the status of other stellar companions whose masses put them in the range between heavy super-Jupiters and lightweight brown dwarfs. Some 'substellar objects' are unquestionably brown dwarfs because studies of the light they give off reveal the spectral fingerprints of core deuterium fusion. In other cases, astronomers must look for other clues to help answer the question: is this a large planet or a brown dwarf?

Researchers have found that giant planets are almost always found orbiting stars that are metal-rich. In astronomical terms, 'metal-rich' means having a relatively high abundance of elements heavier than hydrogen and helium, whether those elements are metals or not. Brown dwarfs, on the other hand, aren't so fussy. This has to do with the different way that planets and brown dwarfs form. Planets grow from the 'bottom up', starting from a small core, rich in heavier elements, which then pulls in more material from its surroundings. Brown dwarfs, by contrast, form in the same way as stars – from dense pockets of a protostellar cloud of gas and dust which collapse under their own weight.

Above a certain mass, believed to be about eighty times that of Jupiter, a newly formed object would become a lightweight red dwarf – the smallest kind of 'normal', hydrogen-burning star – instead of a brown dwarf. The lightest of all known red dwarfs is 2MASS J0523-1403, which is right around the theoretical mass limit for a star and has a diameter a mere 0.09 times the Sun's – smaller than Jupiter. Although only about 40 light-years away it's a million times fainter than the faintest star you can see with your naked eye. Even more diminutive is EBLM J0555-57Ab, which is similar in size to Saturn and is the smallest known red dwarf.[2]

At the other end of the scale are 'sub-brown dwarfs', also known as Y-class brown dwarfs, the smallest, coolest representatives of their species. These form in the same way as other brown dwarfs but overlap with the mass range of super-Jupiters. In fact, the lightest, coolest known brown dwarf, WD 0806-661 B, which is in orbit around a white dwarf, is only seven to nine times as massive as Jupiter and has a surface temperature between 55 and 72 °C.

Although brown dwarfs are made mostly of hydrogen – the lightest element – they can nevertheless be very dense because of the extreme compression of matter inside. CoRoT-3b, for instance, with a mass of around 22 Jupiter masses, is thought to have a density of 26 grams per cubic centimetre, greater than that of osmium, which, as we'll see in Chapter 14, is the densest natural element under normal conditions (22.6 g/cm^3). The surface gravity of CoRoT-3b is correspondingly high – over fifty times that of Earth.

As far as giant planets go, among the biggest is TYC 8998-760-1 b. It's about three times as wide as Jupiter and fourteen times as heavy. Yet, for all its great size, this might-iest of worlds is a youngster. The star it orbits, at a distance slightly more than five times that of Neptune from the Sun, is a mere 17 million years old.[3] Potentially even bigger is HD 100546 b, which, according to some estimates, could be nearly seven times the diameter of Jupiter. But there's a complication because HD 100546 b appears to be a world in the making – a planet that is still condensing from the cloud of gas and dust that surrounds it. Future observations will help decide if it really is the biggest exoplanet known, or if its seemingly immense girth includes, in part, the material in which it's embedded.

CHAPTER 15

Stellar Superstars

Almost every star you can see in the night sky, with your eyes alone, is bigger and more luminous than the Sun. That's just a selection effect: from a great distance, the biggest and brightest in any collection of things will be the ones that stand out. But the fact remains that the Sun is just a run-of-the-mill star – one among many billions of stars of every type in our galaxy. It only seems impressive because we're on its doorstep.

Over time the Sun is getting brighter. It's about 30 per cent brighter now than when it first started to shine 5 billion years ago. Like all stars, during most of its lifetime it makes light and heat by fusing hydrogen into helium in its core. As the helium 'ashes' build up, being more massive than hydrogen, they raise the density of the core causing it to burn hotter – and the Sun to shine brighter. Eventually, all the hydrogen in the core will be used up and gravity, no longer balanced by the outward force of radiation pressure, will squeeze the core smaller, elevating its temperature still more. The extra heat at the centre will ignite hydrogen in a shell surrounding the core and the Sun will enter its twilight years.

As the inert helium core grows, so will the hydrogen-burning shell above it. The Sun's luminosity will rise faster along with the rate at which helium is dumped onto the core. In less than 5 billion years' time, the Sun will be another two-thirds brighter than it is now. And then it will start to swell, even as it becomes yet more luminous. By about 11 billion CE, the Sun will be unrecognisable, with 1,000 times its present-day luminosity and a crimson-hued surface that will have ballooned out monstrously beyond the orbits of Mercury and Venus. The Sun will have been transformed into a red giant.

One factor, above all others, dictates what a star is like and how it will evolve: its mass. At the low end, some stars weigh only about a tenth as much as the Sun. Throughout their long lives, which may run into trillions of years – much longer than the present age of the universe – they eke out their fusion energy reserves very slowly and remain small and cool. 'Red dwarfs' are the commonest type of star in space – twenty of the thirty nearest stars to Earth are of this type.

On the other hand, stars more massive than the Sun live shorter, more spectacular lives. The brightest star in the night sky is Sirius, 8.7 light-years away with double the Sun's mass, 1.7 times its diameter and 25 times its luminosity. Its lifetime as a core-hydrogen burning, 'main-sequence' star will be only about a billion years – less than a tenth that of the Sun.

Over in the constellation Orion are stars on view to the unaided eye, which up close would appear awesome in both size and brilliance. Blue-white Rigel manages to be the seventh brightest star in the night sky even though it lies at least 860 light-years away. As mentioned in Chapter 3, it's classified as a blue supergiant: a massive, very luminous star, with a high surface temperature, that's only recently stopped fusing hydrogen in its core. Rigel is about twenty times as massive as the Sun, with eighty times its diameter and about 120,000 times its luminosity.

Orion is also home to the orange-red star Betelgeuse, another supergiant but with a much cooler, vastly more expansive surface. Betelgeuse is a *red* supergiant so large that if it were put in place of the Sun, its surface would reach out to between the orbits of Mars and Jupiter. Like all supergiants – whether red, blue, white or yellow – Betelgeuse hasn't long to live in cosmic terms. Probably within the next 100,000 years or so, it will explode as a supernova leaving behind a collapsed core in the form of a neutron star or a black hole.

Betelgeuse is huge by solar standards. But it's by no means the largest star known. The biggest that can be seen without binoculars or a telescope is Mu Cephei, also known as the Garnet Star because its deep red hue resembles that of the gemstone. There's some uncertainty about its distance and therefore of its size and brightness but it's definitely bigger than Betelgeuse and would reach out beyond Jupiter's orbit if it replaced the Sun.

Close to the top of the stellar size league is UY Scuti, an immense star that lies between 9,000 and 10,000 light-years from Earth. It's described as either an extreme red supergiant or a red hypergiant and has a diameter about 1,700 times that of the Sun. Five billion Suns could be crammed inside its stupendous bulk. In fact, UY Scuti is so large that it exceeds the theoretical size limit for stars – around 1,500 solar diameters – conventionally accepted by astronomers. There's an ongoing debate about how the biggest known stars formed in the first place and how they remain stable for most of their brief lives.

In the number one slot, based on size, is a star known as Stephenson 2-18 (abbreviated to St 2-18), which lies on the outskirts of a cluster of stars 19,000 light-years away.[1] Its diameter is estimated to be 2,150 times that of the Sun, or 20 times the Earth–Sun distance. A ray of light, travelling at 300,000 kilometres per second, would take nearly nine hours to go around its surface, compared to 14.5 seconds for the Sun. If placed at the centre of the Solar System, the surface of St 2-18 would engulf the orbit of Saturn.

Stephenson 2-18 is also extreme in another way: it's fantastically bright – as luminous as a third of a million Suns. The only reason it doesn't dazzle us with its brilliance is that it's so far away. Stars can appear bright in the night sky because they're relatively nearby. The third brightest, after Sirius and Canopus, is Alpha Centauri (actually a three-star system), which is less than 4.5 light-years away. But the most luminous of stars, in absolute terms, are those which radiate prodigious amounts of light from their surfaces.

Among stars that can be seen with the naked eye the most intrinsically luminous is probably Zeta[1] Aurigae – a white hypergiant that dwells within a young cluster of hot, bright stars known as the Scorpius OB-1 association. It shines as brightly as 850,000 Suns.

Further up the league of stellar brilliance are a number of members of the Arches Cluster – the densest known star cluster in our galaxy, located about 100 light-years from the centre of the Milky Way and 25,000 light-years from Earth. The most prominent inhabitants of the Arches are thirteen so-called Wolf–Rayet stars and eight O-type hypergiants. Both these types of star are extremely hot and massive and dominate the top positions in our luminosity league. Currently occupying first position is R136a1, which lies in the Tarantula Nebula – a vast region of hot gas and young stars situated within a satellite galaxy of ours, the Large Magellanic cloud. R136a1 is a Wolf–Rayet star with 7 million times the luminosity of the Sun. It's also one of the heaviest known stars at 222 solar masses. At the number two and three slots are its neighbours in the Tarantula Nebula, R136a2 and BAT98-99, again both supermassive, superhot Wolf–Rayets.[2]

The brightest known stars, including the three just mentioned, are also among the most massive. All are close to, or more than, 200 times as heavy as the Sun. It's possible that in the very early days of the universe, shortly after the first stars formed, some stellar behemoths were even heavier. Among the so-called 'Population III' of the dawn cosmos there were perhaps stars weighing up to several hundred, or even 1,000, solar masses, and with corresponding colossal luminosities. No such stars have yet been observed but there is some indirect evidence for them and, with a new generation of very powerful telescopes coming on stream, it may not be long before we catch our first glimpse of these extreme superstars.

CHAPTER 16

Big News

BACK IN THE third century BCE, the Greek mathematician Archimedes wrote a book called *The Sand Reckoner* in which he tried to figure out how many grains of sand would fit into the entire universe. To do that he had to estimate the size of the universe based on the best information then available. The figure he came up with for the maximum diameter of all of space was 100 trillion stadia or, in modern terms, about 2 light-years.

A light-year is a ridiculously big unit. It's how far light, moving at 300,000 kilometres per second – the highest speed possible – travels in a year: 9.5 trillion kilometres. For ancient thinkers to contemplate something twice that far across is pretty impressive. But today we have to stretch our imagination to deal with things that are vastly larger.

The galaxy we live in, the Milky Way, is a community of several hundred billion stars spanning about 90,000 light-years. In the early 1920s, a debate came to a head about the nature of the universe: whether our galaxy essentially *was* the entire universe, or whether certain spiral and elliptical 'nebulae' were, in fact, other galaxies, or 'island universes' as they were called at the time. It soon became clear that the Milky Way isn't alone. The Great Spiral in Andromeda turns out to be a spiral galaxy even bigger than our own and lying 2.5 million light-years away.

We can just make out the Andromeda Galaxy, as a fuzzy patch in the night sky, with our unaided eyes. The light we're receiving from it now started its journey before the earliest human ancestor, *Homo habilis*, appeared, and when members of the genus *Australopithecus* still walked the Earth. Yet, it turns out that Andromeda is one of our nearest galactic neighbours. The Milky Way and the Andromeda spiral are the dominant members of the Local Group – a gravitationally bound cluster of about eighty galaxies, most of them 'dwarfs', occupying a volume of space roughly 10 million light-years across.

Just as most stars are part of galaxies, so most galaxies are found in clusters. Nearest to the Local Group is the M81 Group, a cluster of about three dozen galaxies, also with two outstanding members, the spiral M81 and the starburst galaxy M82, in which new stars are forming at a ferocious rate. But small groups like these are minnows in the wider universe. About 65 million light-years from us lies the Virgo Cluster, between 1,300 and 2,000 galaxies strong and featuring some giant elliptical systems, including M87, one of the largest and most massive galaxies in our neck of space. M87 has a supermassive black hole in its core, which became the first such object to be directly imaged. In 2021, astronomers revealed pictures made of M87's dark centre and its environs built up using observations by the Event Horizon Telescope.

Continuing up the hierarchy, most clusters of galaxies belong to clusters of clusters called superclusters. The Virgo Cluster sits at the heart of the Local Supercluster, to which the Local Group and scores of other clusters also belong, with a total population of about 47,000 galaxies. The Local Supercluster is thought to have a diameter of about 110 million light-years and a mass roughly 1,200 trillion times that of the Sun. It's impossible to imagine how something could be so big: more than a thousand times wider than the Milky Way, which itself is of breathtaking size.

Just as there are other clusters of galaxies beyond the Local Group, so there are other superclusters. These are often named after the part of the sky in which they lie as seen from Earth. Among the nearest to the Local Supercluster are the Coma Cluster, and others named after the constellations Hydra, Centaurus and Perseus–Pisces. The last of these sprawls across almost 300 million light-years of space. In the time it takes a beam of light to travel from one end of the Perseus–Pisces Supercluster to the other, life on Earth went from the Carboniferous coal-forming swamps, inhabited by amphibious ancestors of the dinosaurs, to the modern day.

And yet in a quest for the biggest things in the universe we're not done. Recently, astronomers have found that our Local Supercluster is just part of an even larger structure that's been named the Laniakea Supercluster – 'Laniakea' meaning 'immense heaven' in Hawaiian.[1] Home to around 100,000 galaxies stretched over 520 million light-years, it has four main parts: our own supercluster and those of the Hydra–Centaurus, Pavo–Indus and Southern Superclusters. With this new knowledge, it's clear that the Laniakea Supercluster really *is* our local supercluster and that the movements of our own modest Local Group are influenced by a wider assemblage of galaxies than was previously supposed.

One of the mysteries of astronomy is a phenomenon called the Great Attractor. We're not sure exactly what it is but we do know that the pull of its gravity is influencing the motion of the Milky Way and all of the other galaxies in our neighbourhood. It now seems that the Great Attractor is Laniakea's gravitational focal point, but its nature remains uncertain because the Attractor lies in the Zone of Avoidance – an area of the sky that's partially obscured by all the gas and dust lying in the plane of our galaxy.

We've come a long way from the 2-light-year-wide cosmos of Archimedes to a realisation of the true vastness of our home supercluster. But even Laniakea isn't the largest known structure in the universe. The Saraswati Supercluster, 4 billion light-years distant, was discovered by Indian researchers in 2017 and has a maximum length from one end to the other of about 650 million light-years.[2] And still larger things have come into focus as astronomers have mapped the distribution of matter over larger and larger scales.

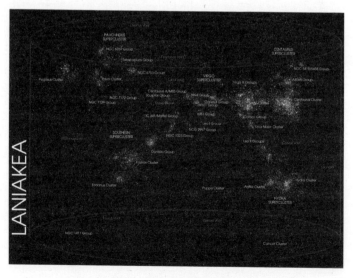

A map of the Laniakea Supercluster and its
component galaxy clusters.

Laniakea itself is part of the Pisces–Cetus Supercluster Complex, which is a vast type of entity known as a galaxy *filament* or *wall*. The universe, it turns out, when viewed over distances of billions of light-years, is seen to consist of walls of gravitationally bound superclusters surrounding immense empty spaces called voids. The Pisces–Cetus Supercluster Complex is estimated to be about 1 billion light-years long and 150 million light-years wide. It's so massive that the Virgo Supercluster – what we used to think of as our local supercluster – accounts for only about a thousandth of its mass. The Sloan Great Wall is larger with a length of 1.37 billion light-years, putting it on a par with the South Pole Wall found as recently as 2020. It may seem surprising that something so large could go unnoticed by astronomers for so long. But, as in the case of the Great Attractor, our view of this structure is compromised by the fact that it lies in a direction along the plane of our galaxy.

Between the filaments are voids, which also have names even though they're essentially comparatively empty. The closest one to us is the Keenan, Barger and Cowie (KBC) void, with a diameter of 2 billion light-years.

Laying claim to be the largest known cosmic structure is the Hercules–Corona Borealis Great Wall, a colossal distant filament stretching up to 10 billion light-years across – more than 10 per cent of the diameter of the observable universe. What could possibly be bigger?[3]

The entirety of all the interconnected galactic filaments and voids across space is referred to as the cosmic web. If this can be regarded as a single object, then there can be nothing bigger than it anywhere inside the part of the universe that we can see. The cosmic web stretches across all of the observable universe, a ball-shaped region with a diameter of about 93 billion light-years. We can't look beyond the edge of the observable universe because light hasn't had enough time since the Big Bang to reach us from further away.

For now, the Hercules–Corona Borealis Great Wall is probably the best contender for the title 'single biggest thing'. But its very existence creates a headache for astronomers. Although matter is obviously lumpy at the scale of galaxies, clusters of galaxies and even superclusters, current cosmology suggests that the universe ought to be relatively smooth overall. Apparently, though, if the Hercules–Corona Borealis Great Wall is real, there can be lumpiness over a distance of at least a tenth of the diameter of the observable universe. Theorists have a job on their hands to explain how that can be so.

CHAPTER 17

Far, Far Away

IN SEPTEMBER 1966, astronauts Charles 'Pete' Conrad and Richard Gordon, aboard their Gemini 11 capsule, fired the engine of the Agena Target Vehicle with which they'd docked. The rocket burn raised the maximum height of their orbit to 1,373 kilometres – the furthest anyone, at that time, had travelled from Earth. It remains the highest orbit around our planet ever attained by a crewed spacecraft.

The greatest distance that humans have travelled from Earth is about 400,000 kilometres, by the eight Apollo missions that either landed on or just circled around the Moon between 1968 and 1972. However, some of our robot emissaries have ventured much farther afield. On a barren, orange, rock-strewn moon of Saturn sits the now-defunct *Huygens* spacecraft. After it parachuted down to the surface of Titan in January 2005 it achieved the farthest landing from Earth that a spacecraft has ever made and the only touchdown to date on a body in the outer Solar System.

A number of space probes, though, have flown much farther afield. Five, in fact, are heading out of the Solar System altogether: *Pioneers 10* and *11*, *Voyagers 1* and *2*, and *New Horizons*. Of these, currently the farthest from the Sun – and the most distant human-made object – is *Voyager 1*, at a distance of about 24 billion kilometres or roughly 160 AU (where 1 AU, or astronomical unit, is the average Earth–Sun distance). It's so far away that radio signals from it, moving at the speed of light, take nearly a day to reach us.

Both of the *Voyager* probes are now farther from the Sun than the most distant natural object in the Solar System seen to date through a telescope. That title is currently held by 2018 AG_{37}, an icy-rocky body, a few hundred kilometres across. Appropriately nicknamed 'Farfarout', it takes nearly a millennium to go once around its very elongated orbit, ranging between about 175 and 27 AU from the Sun.

Astronomers suspect there are many billions of objects like 'Farfarout' orbiting the Sun that are too small and remote for us to see at present. The most distant objects of all in the Solar System are thought to be those in the Oort cloud. This, theory suggests, is a gigantic region home to countless frozen bodies, some of which are occasionally deflected onto paths that carry them into the inner Solar System as long-period comets. The outer part of the Oort cloud may be anywhere from 10,000 to 100,000 AU from the Sun – up to a third of the distance to the nearest star.

Now that Pluto has been demoted to 'dwarf' status the outermost planet of the Solar System is Neptune, which averages 30 AU, or 4.5 billion kilometres, from the Sun. But there are many more planets going around other stars. Most of the exoplanets that have been discovered over the past thirty years or so lie within a few hundred light-years of Earth. Planets are much smaller and dimmer than the stars they go around, so naturally the closer ones are easier to find. But a few exoplanets are known within the Milky Way that lie at much greater distances.

In 2006, a search for exoplanets was carried out using the Hubble Space Telescope. Called SWEEPS (Sagittarius Window Eclipsing Extrasolar Planet Search) it used the transit method, in which the strategy is to look for slight dips in the amount of light we receive from a star caused by planets passing in front of the star's disc as they move in their orbits. The Sagittarius Window is a region of the sky, looking in the direction of our galaxy's central bulge, where the view of stars located in the bulge some 27,000 light-years away is relatively unobscured by clouds of gas and dust. Two Jupiter-size planets, SWEEPS-04 and SWEEPS-11, were found orbiting a star located 27,710 light-years from the Sun. That puts them further away than the galactic centre – but they're not the most distant exoplanets known within the Milky Way galaxy.

The bright spiral arms of our galaxy lie in a flattened disc that spans just under 100,000 light-years. But the Milky Way extends well beyond this showy, star-rich region. Its outer halo is a sparse spherical volume that stretches out at least half a million light-years from the galactic centre. Among the halo population are the most distant stars that are still gravitationally bound to the Milky Way. Astronomers hunt for these remote suns not simply because they're curiosities but because they can give us valuable clues about the formation and evolution of our home galaxy. Using the Multiple Mirror Telescope in Arizona, astronomers studied two remote red giants called ULAS J0744+25 and ULAS J0015+01 and found that they lie at distances of 775,000 and 900,000 light-years, respectively. That's more than 50 per cent farther from the Sun than any other known star in the Milky Way and about one third of the distance to the Andromeda Galaxy.[1]

Considering how common planets appear to be around nearby stars, it seems inevitable that there must be vast numbers of them throughout the universe – both within galaxies and roaming free, perhaps unmoored from the stars they once orbited, in the intergalactic void. There have been a few claims of detection of extragalactic planets, but these have yet to be confirmed.

In 2009, researchers recorded a microlensing event in our nearest neighbouring large galaxy, Andromeda.[2] Microlensing happens when the gravitational field of an unseen foreground object focuses the light from a more distant source, such as a star or galaxy, rendering it visible to us here on Earth. The details of the Andromeda event, known as PA-99-M2, fit with a scenario in which the lensing object is a star accompanied by a planet with a mass of about six Jupiters. Something similar happened in 1996 when a team of astronomers spotted an unusual fluctuation in the light coming from a remote quasar – the incredibly bright nucleus of an active galaxy. The light changes matched those that might have been caused by a planet roughly three times as massive as Earth, lying within the galaxy responsible for the lensing. That galaxy is about 4 billion light-years away, which would make the planet easily the most distant ever found. The trouble is, the chance alignment that led to the discovery will never happen again, so we shall never know for sure.

Microlensing was also at work in revealing the most distant star ever seen. Officially known as WHL0137-LS, but also named Earendel from the Old English for 'morning star' and with a nod to Tolkien, this ancient stellar giant was detected by chance when a galaxy cluster aligned with it and magnified its light thousands of times.[3] Earendel shone millions of times more brightly than the Sun when the universe was less than a billion years old. We're seeing it now as it was about 13 billion years ago, although in the time it's taken its light to reach us the universe has expanded so much that Earendel is now a staggering 28 billion light-years away. Given its brilliance and presumed enormous mass, it exploded long ago.

It's a well-known but remarkable fact that the farther we look out into space the further we look back in time. The most distant objects we've ever seen existed not long after the Big Bang itself when the universe came into being. By studying them, astronomers can learn what the first galaxies were like and how they may have formed from primordial material in the dawn cosmos.

In 2016, researchers announced that, using results from the Hubble Space Telescope and the Spitzer Space Telescope, which works at infrared wavelengths, they'd found what was then the most remote galaxy known. Called GN-z11 we're seeing it as it was when the universe was just 400 million years old. Compared with the Milky Way, GN-z11 is about 4 per cent of the size, has 1 per cent of the mass, and is forming new stars about twenty times as fast.

In their search for the earliest galaxies, astronomers now have a powerful new tool in the James Webb Space Telescope (JWST) – the largest instrument of its kind ever launched, which began scientific observations in July 2022. Already it's led to the discovery of several galaxies that appear to be even older, and therefore farther away, than GN-z11. Topping the list is JADES-GS-z13-0 (JADES is an acronym of JWST Advanced Deep Extragalactic Survey), which we're looking at as it was a mere 325 million years after the Big Bang.[4] Its light has taken 13.5 billion years to reach Earth. Allowing for cosmic expansion while that light has been on its way to us, the present distance to JADES-GS-z13-0 is 33.6 billion light-years. That makes it the most distant object ever seen and poses a challenge for scientists – how to explain the formation of stars and galaxies at such an early stage.

CHAPTER 18

Ka-boom!

EXPLOSIONS ARE AS old as – well, the universe itself. Human-made ones are a little more recent and modest in scale. The first chemical explosive to be concocted was black powder, also known as gunpowder – a mixture of charcoal (a form of carbon), sulphur and potassium nitrate or saltpetre. Charcoal is the fuel; potassium nitrate contains the oxygen for combustion; and sulphur lowers the temperature needed to trigger the reaction. The key to making good gunpowder is to grind the mixture up finely. This allows the components to be in close contact for a speedier reaction – and the faster the reaction, the bigger the bang.

Several recipes for inflammable mixtures were written down by Chinese alchemists as early as the fourth and fifth centuries, and a Taoist book from 850 CE warns of three specific formulations that are too dangerous for experimentation. This date agrees with what appears to be the first use of gunpowder in military applications in the late Tang Dynasty. Thereafter, during the Song and Yuan dynasties (960–1368), it's clear that the Chinese had developed various ingenious means of deploying the explosive against their enemies, including catapults, rockets, 'fire cannons' and 'fireballs'. The method of making gunpowder spread first to the Arab world and then to

Europe, where by the 1350s it had become an effective weapon on the battlefield.

Any chemical reaction that can be sped up enough has the potential to be explosive – even burning something as seemingly innocuous as flour. Just about any kind of carbohydrate dust, including sugar, pudding mix and fine sawdust, will explode once ignited. The grains are so tiny that they burn instantly. When one grain burns, it lights other grains nearby, and the flame front can flash through a dust cloud with explosive force.

The first documented dust explosion happened on December 14, 1785, in Mr Giacomelli's bakery warehouse, in Turin, Italy. We know exactly what happened because of a detailed reconstruction by Count Morozzo in the *Memoirs of the Academy of Science of Turin*. An oil lamp, mounted to help flour workers see in the evening, ignited dust thrown up by normal handling operations. Then, writes the Count:

> [A]n explosion threw down the windows and window-frames of the shop, which looked into the street; the noise was as loud as that of a large cracker, and was heard at a considerable distance... [T]he inflammation proceeded from the flour warehouse ... where a boy was employed in stirring some flour. The boy had his face and arms scorched by the explosion; his hair was burned, and it was more than a fortnight before his burns were healed.

In 1878, a far more serious explosion of grain dust destroyed the Washburn 'A' Mill in Minneapolis, Minnesota – the largest industrial building in the city and the largest grain mill in the world. Twenty-two people died in the ensuing fireball and the blast was heard 10 miles away in St Paul.

By the middle of the nineteenth century, chemists had started to develop explosives much more powerful than gunpowder. Nitroglycerine, first synthesised by the Italian chemist Ascanio Sobrero in 1847, is an oily liquid that explodes twenty-five times faster than gunpowder and with three times the energy. The problem is, on its own or mixed with gunpowder, it's terrifyingly unstable.

In April 1866, three crates of nitroglycerine were shipped to California for the Central Pacific Railroad, which wanted to test its effectiveness for blasting a tunnel through the Sierra Nevada Mountains. But en route, one of the crates exploded at a Wells Fargo office, destroying the building and killing fifteen people. Swedish chemist Alfred Nobel, owner of one of the largest explosives factories in Europe, lost his brother to a nitroglycerine explosion. That accident, and government bans on the use of the explosive in the US, the UK and elsewhere, encouraged Nobel to find a safer alternative. He found it in dynamite: a stable explosive made by mixing nitroglycerine with a type of clay known as kieselguhr.

Other explosives were soon developed, including gelignite, TNT (trinitrotoluene) and picric acid. The explosive yield of a bomb or other type of explosion is often expressed in terms of tons, kilotons or megatons of TNT. On August 4, 2020, one of the most powerful chemical explosions in recent times rocked the port of Beirut and surrounding area when 2,750 tons of improperly stored ammonium nitrate was accidentally detonated. The blast, equivalent to about 500 tons of TNT, caused 190 deaths and more than 6,000 injuries.

On December 6, 1917, just offshore in Halifax, Nova Scotia, two ships collided, one of which, the SS *Mont-Blanc*, was fully loaded with TNT, picric acid, the highly flammable fuel benzol and guncotton. In the ensuing explosion, over 1,600 people were killed instantly, buildings across an area of 1.6 square kilometres were flattened, and *Mont-Blanc*'s forward 90-mm gun was hurled 5.6 kilometres through the air. Equivalent to 3 kilotons of TNT, the Halifax explosion was the biggest chemical one in history – but far from the biggest blast caused by human hand.

Nuclear weapons can unleash vast amounts of energy. The bomb dropped on Hiroshima, which was small by today's standards, had a yield equivalent to about 67 kilotons of TNT. It was also extremely inefficient – only 1.7% of the 64 kilograms of uranium-235 it carried actually released destructive energy. The most powerful nuclear device ever detonated was the Tsar Bomba, also known as Big Ivan, which was tested by the Soviet Union in 1961.[1] It had a yield equivalent to 50 megatons of TNT – the same as 3,800 Hiroshima bombs going off at once. Dropped by parachute from a plane over the deserted island of Novaya Zemla, it exploded 4 kilometres above the ground, causing a flash that could be seen 1,000 kilometres away and producing a towering mushroom cloud that rose more than 60 kilometres high.

Natural explosions on Earth, such as those caused by large volcanic eruptions and asteroid strikes, can dwarf even the most destructive forces unleashed by man. But they, in turn, are made to look insignificant by some of the colossal blowouts that occur in space. Explosions on the surface of the Sun, known as solar flares, typically release a million times more energy than an erupting volcano – and the Sun, by astronomical standards, is a quiet, unassuming sort of star.

When larger, more massive stars get to the end of their lives, they're anything but quiet. Stars that weigh more than ten times the mass of the Sun undergo a sudden, startling transition when all of their useful fusion fuel runs out. No longer able to buoy themselves up against the inward press of gravity by making new light and heat, they suddenly collapse and then instantly rebound, hurling most of their contents into space at speeds up to an eighth that of light. The violent demise and huge outpouring of energy of a supernova enables it to outshine an entire galaxy of hundreds of billions of stars for a few days or weeks. In that time, its output can match that of an average star, like the Sun, over its entire 10-billion-year lifetime.

What blast could be bigger than an entire star bursting apart in a fraction of a second? The answer: an entire humongous star – weighing up to 100 times as much as the Sun – bursting apart *and* firing out a ferocious volley of high-energy gamma rays. Such mega blowouts are thought to be a principle cause of rare events known as gamma ray bursts (GRBs). These can unleash 10 to 100 times as much energy as a conventional supernova and be detectable across billions of light-years.

But just when astronomers thought that the most massive star detonations must surely top the hit parade of cosmic explosive events, along comes something even bigger and more violent. It's been dubbed BOAT – 'brightest of all time' – because of its extraordinary violence and luminosity. BOAT was discovered by combining the data from several instruments: in space, NASA's Chandra X-ray Observatory and the European Space Agency's XMM-Newton, and on the ground, the Murchison Widefield Array and the Giant Metrewave Telescope. The origin of the explosion lay within the Ophiuchus cluster of galaxies, some 390 million light-years from Earth.

Researchers investigating the Ophiuchus cluster found an enormous bubble in the hot gas that pervades the space between its galaxies.[2] This intergalactic cavity is 1.5 million light-years across – wide enough to fit fifteen Milky Way galaxies in a row. It was evidently caused by jets firing away from the central supermassive black hole of a large galaxy at the heart of the cluster. But what astonished scientists was the energy involved in the phenomenon – a billion times the energy of a powerful supernova explosion, released, not in a sudden, short burst, but in a continuous vast outpouring that lasted for months and possibly years.

CHAPTER 19

Racing Through the Cosmos

THE SPACECRAFT THAT circle Earth fastest are those in the lowest orbits. The International Space Station, at a mere 420 kilometres (260 miles) above the surface, moves at about 28,000 kilometres per hour (17,500 miles per hour). If a spacecraft orbits much lower than this it'll skim the outer layers of the atmosphere, the frictional drag of which will slow it down and eventually cause it to re-enter. So 18,000 mph is about tops for Earth-orbiting satellites.

To escape Earth's gravitational pull altogether, a spacecraft has to start off with a speed of 11.2 kilometres per second (7 miles per second). That's about 40,000 km/h (25,000 mph). The fastest that humans have ever travelled in space is just below Earth-escape velocity in the case of the Apollo 10 astronauts when, upon their return from orbiting the Moon, their capsule reached 39,897 km/h.

The highest speed of re-entry into Earth's atmosphere was 46,100 km/h by the *Stardust* probe on its return, in 2006, from Comet Wild 2 with samples of cosmic dust. The highest speed of escape from Earth was 58,500 km/h, in the same year, by *New Horizons* as it departed on its way to Pluto and beyond.

New Horizons is one of five spacecraft that are leaving the Solar System altogether. In fact, two of them at least, *Voyager 1* and *Voyager 2*, have already reached interstellar space. To be sure of breaking free from the Sun's gravitational pull, without any further boost to its speed, an object would have to be travelling at about 151,300 km/h (94,000 mph) when it left Earth orbit. None of our current star-bound probes did that. The reason they're now escaping the Solar System is that their speed was boosted by intentionally passing close by planets along the way and thereby receiving gravitational assists.

When we're talking about the speed of things in space it's important to be clear about what the speed is being measured relative to: Earth, the Moon, the Sun, or some other body. The *Galileo* space probe was travelling at more than 173,000 km/h relative to Jupiter when it entered orbit around the giant planet in 2003. Thirteen years later, another robotic spacecraft, *Juno*, went into Jupiter orbit at 209,000 km/h.

At *Juno*'s speed, the journey from London to New York would take two minutes and a trip to the Moon a little under two hours. But much faster still is the current spacecraft speed record holder: the Parker Solar Probe.[1] Launched in 2018, its mission is to study the Sun's outer corona from an incredibly small orbit. In 2021, it reached a speed of 587,000 km/h relative to the Sun's surface. In 2025, it will approach the Sun to within 6.9 million kilometres, less 10 solar radii – at which point it will be travelling at 690,000 km/h (430,000 mph).

Astonishingly fast though this may sound, it's only 0.064 per cent of the speed of light. If the Parker Solar Probe were to travel directly towards the nearest star beyond the Sun, Proxima Centauri, 4¼ light-years away, at its maximum speed, it would take it over six and a half thousand years to get there.

In decades to come we might build faster spacecraft, powered by such means as nuclear fusion. Craft like this could travel interstellar distances in much shorter times. But until then we have to look to the natural universe for things that travel at exceptional speed.

**Artist's impression of the Parker Solar Probe
approaching the Sun.**

Within the Solar System, the fastest moving objects relative
to the Sun – following the same principle as Earth-orbiting
satellites – are those with the smallest orbits around the
Sun. Mercury, the innermost planet, has an orbital speed of
47 kilometres per second (169,200 km/h). Not surprisingly,
this is slower than the Parker Solar Probe because the latter
moves well inside Mercury's orbit.

The real speedsters of the Solar System are Sun-grazing
comets, an extreme example of which was the Great Comet
of 1843, which came within 121,000 kilometres of the Sun's
surface. At the point of closest approach it would have been
travelling at about 570 km/s or about 2 million km/h. Most
comets that come this close to the Sun disintegrate, but the
Great Comet of 1843 avoided this fate and is now headed
back into the remote depths of the Solar System, more than
five times farther out than Neptune.

All stars in our galaxy are in motion relative to the centre of the Milky Way. The Sun is travelling at about 24 km/s, or 864,000 km/h, around the galactic centre. At this speed it takes 230 million years or so to complete one circuit of the galaxy – a time known as a galactic year. Generally, stars that are closer to the centre move faster, but there are exceptions.

So-called runaway stars have experienced some traumatic event that has hurled them away from their old paths through space at a tremendous rate. If a star has a neighbour that explodes as a supernova, or if it passes very close to another star, it may be thrown onto a new course at high velocity. A famous example involves three runaway stars – AE Aurigae, 53 Arietis and Mu Columbae – all of which are flying away from each other at speeds of over 100 kilometres per second. Tracing back their motions, astronomers have found that their paths intersect at a point near the Orion Nebula about 2 million years ago. In this location today is a giant structure known as Barnard's Loop – the remains of a supernova, which launched the stellar trio onto their present trajectories at a time when our *Homo habilis* ancestors walked the Earth.

Faster still are hypervelocity stars, some of which may exceed the escape velocity of the galaxy. A typical hypervelocity star moves faster than about 1,000 kilometres per second relative to the galactic centre. The fastest recorded to date – and the fastest-moving star on record – is S4714, which orbits closely around the supermassive black hole at the galaxy's heart. It's one of hundreds of stars near to the galactic centre, which are being whipped around their orbits at breakneck speed by the fierce gravitational pull of the monster black hole. S4714 is careering around at 24,000 km/s – more than 8 per cent of the speed of light.[2]

Even faster hypervelocity stars may be awaiting discovery. Astronomers estimate that if a star gets too close to the central black hole, it may suffer one of two drastic fates: be torn apart by the intense gravitational field or slung out of the Milky Way and into the intergalactic void. Hundreds of hypervelocity stars have been found over the past few years – most of them on escape trajectories from the galaxy.

A star needs a speed of at least 550 km/s to break free from its galactic moorings. Of these stellar escapees, the current record holder is a hot, white star called S5-HVS1, 29,000 light-years from Earth in the southern constellation Grus, which is heading for intergalactic space at 1,755 km/s (nearly 4 million mph). But astronomers suspect there may be much faster stars departing the Milky Way, having had a close encounter with the central supermassive black hole and been slung out at speeds of between 30,000 and 100,000 km/s – between a tenth and a third of the speed of light.

Stars move relative to the galaxies they inhabit and galaxies move relative to each other. Our own great city of stars, the Milky Way, and its nearest large neighbour, the Andromeda Galaxy, are approaching each other at about 110 km/s and will, in fact, collide in about 5 billion years' time. Most galaxies, however, are moving away from us due to the overall expansion of the universe. The most distant ones known, which we're seeing as they were when the universe was just a few hundred million years old, are flying away from us at well over nine-tenths the speed of light because of cosmic expansion.

Other near-light-speed phenomena involve less substantial pieces of matter than stars and galaxies. Jupiter-sized blobs of hot gas have been found hurtling away from remote hyperactive galaxies, known as blazars, at 99.9 per cent of the speed of light. The amount of energy needed to make these plasma blobs travel so fast is astonishing. All the energy we produce on Earth in a week could barely accelerate a dollop of plasma with the mass of a bowling ball to such speeds.

Nothing travels faster than light. Anything that has even the slightest amount of mass when not moving can never quite reach light-speed no matter how much energy is expended in the effort. But some things come very, very close. The fastest things in the universe, apart from light itself, are ultra-high-energy cosmic rays. These charged particles – mostly protons – can attain not just 99.9 per cent light-speed but 99.9 followed by 19 more nines per cent. How they have attained such fantastic speeds and energies remains an open question, but high on the suspect list are collisions between clusters of galaxies and the accelerating power of supermassive black holes at the fearsome heart of blazars.

MATERIALS

CHAPTER 20

How Dense Can You Get?

'As heavy as lead'. It's a common expression and it's true – lead *is* heavy for its size. A cricket ball weighs about 160 grams and has a radius of about 3.6 centimetres. A solid ball of lead the same size would weigh 2.2 kilograms. Swap the lead for gold, though, and the ball would be more than half as heavy again at close to 3.8 kilograms.

Dividing an object's mass by its volume tells you its density. The density of lead is 11.3 grams per cubic centimetre, that of gold 19.4 grams per cubic centimetre. Denser still are a handful of other elements, including tungsten and platinum. But densest of all is a precious metal that's much rarer than either gold or platinum: element 76, osmium, which we first encountered in Chapter 14. A cricket ball made of osmium would weigh about 4.4 kilograms – slightly more than the ball used in women's shot put. No substance on Earth has a greater density.[1]

Extremes on Earth, however, pale into insignificance compared with those found in other parts of the cosmos. An inkling of what might be possible comes from thinking about the make-up of familiar substances. Everything around us is composed of atoms. An atom of osmium, for instance, consists of a nucleus containing 76 protons and (most commonly) 116 neutrons around which circulates a cloud of 76 electrons.

Only the nucleus is substantial, accounting for almost all the atom's mass. And yet the nucleus is tiny compared with the atom as a whole. An osmium nucleus has a radius of just 7.2×10^{-15} metres, or 7.2 picometres. That's about 25,000 times less than the radius of an osmium atom, equivalent to an atom-to-nucleus volume ratio of about 15 trillion.

Incidentally, you may be wondering why osmium has a higher density than the element with the heaviest nucleus found in nature, uranium. The reason is that an osmium atom is relatively small in relation to the mass of its nucleus because the outermost electrons are pulled in tightly, resulting in an overall compact size.

Most of an atom is empty space. So, what outwardly may seem like a hard, solid object, such as a rock or an iron bar, is really almost ghostlike in its internal emptiness. Despite that, it's very difficult to make something that's solid any denser by squeezing it because atoms are formidably strong and rigid. Even the immense pressures found deep within a planet can only compress solid matter to a modest degree.

Stars, on the other hand, are much more massive than planets, so that their deep interiors are squeezed far harder by the great weight of the overlying layers. The pressure at the centre of the Sun is about 265 billion times greater than the atmospheric pressure at Earth's surface. What's more, the Sun's core is extremely hot – about 15,000,000 °C . At this temperature, atoms are stripped of all their electrons, leaving behind bare nuclei. The hot soup of free electrons and naked nuclei is known as a plasma – a fourth state of matter that is different from solid, liquid or gas.

Because the atomic structure has been broken down in a plasma, it's possible to squash the constituent electrons and nuclei much closer together than would be the case with ordinary matter, resulting in a higher density. This is true even though the nuclei present in the core of most stars, such as the Sun, are those of the lightest elements, hydrogen and helium. The density in the Sun's core is estimated to be as high as 160 grams per cubic centimetre – seven times higher than the density of osmium. A cricket ball made of hot plasma scooped from the centre of the Sun would weigh about 31 kilograms, or roughly as much as a ten-year-old child.

In the distant future, the density deep inside the Sun will rise further. At some point, several billion years from now, all the hydrogen in the Sun's core will have been converted into helium by nuclear fusion. To begin with, this helium won't be hot enough to fuse into heavier elements. Instead, it will be squeezed harder and harder by the weight of the overlying layers, causing its density to soar to about 10 million grams – 10 tonnes – per cubic centimetre. Only then will another force come into play that prevents further compression. The electrons in the compressed helium plasma will resist being pushed closer together by a phenomenon called the Pauli exclusion principle. According to this rule, no two neighbouring electrons can be in exactly the same state as defined by four special quantities known as quantum numbers.

Matter squashed so hard that the exclusion principle resists further compression is said to be 'degenerate'. Later in their evolutionary journey, Sun-like stars grow to become red giants before casting off their bloated outer layers to leave behind hot, planet-sized white dwarfs. These exposed stellar cores are made entirely of degenerate matter, doomed to cool for eternity, but supported against further gravitational implosion by the pressure of electrons resisting further overcrowding.

The Sun is a very ordinary star, which looks spectacularly big and bright only because it's so close. Many stars, as we've seen, are bigger, brighter and, more to the point, more massive than the Sun. Their endpoints are correspondingly more extreme. A star much heavier than the Sun explodes violently at the end of its life leaving behind a core that can't stabilise as a white dwarf. Instead, if the remnant core weighs more than about 1.4 times as much as the Sun, electron degeneracy isn't enough to prevent the core from collapsing further under the force of its own gravity. Electrons and protons are crammed together so hard they combine to become neutrons, giving rise to a miniature but massive stellar core known as a neutron star.

Imagine the mass of one and a half Suns crammed into a ball just 10 kilometres wide. The stuff of which it's made is 'neutronium' – an ultra-dense form of matter in which neutrons, perhaps with a few protons and electrons sprinkled here and there, are packed tightly together. A neutronium cricket ball would weigh about 200 billion tonnes, or roughly the same as Mount Everest. The only thing stopping a neutron star from squeezing itself even smaller is neutron degeneracy due to the exclusion principle acting on closely packed neutrons.[2]

Conceivably, there's a more dense state that may exist in the cores of some neutron stars. If the temperature and pressure in the heart of a neutron star are high enough, it's hypothesised, the degeneracy pressure of the neutrons might be overcome. If this were to happen, the neutrons would be forced to merge and break apart into quarks – the building blocks of particles such as protons and neutrons. This would lead to an ultra-dense phase of matter called quark matter.[3]

It's an open question whether quark matter and quark stars exist in the universe today. But what's certain is that if the stellar core left behind after a star explodes has a mass greater than about twice that of the Sun, then there's nothing that can stop it from progressing to the ultimate state of gravitational collapse. Above the limit of two solar masses, not even the degenerate pressure of electrons, neutrons or quarks can prevent gravity from crushing the core, in the blink of an eye, into a black hole.

Present-day physics has only a tenuous grasp of what goes on inside a black hole – within its event horizon from which not even light can escape. If the black hole doesn't rotate, current theory predicts that the density of matter at its centre is infinitely high. That prediction reflects our ignorance rather than actual fact: the laws of physics as we know them break down at the central, so-called singularity. But it's a different story if the black hole rotates, as in reality it almost certainly would because it evolved from a star which itself spun around. The singularity inside a rotating black hole, into which all its mass is concentrated, takes the form not of a point but of a ring-like shape. The density of its matter, therefore, although incredibly high, will at least be finite.

Black holes are the holders of numerous physical records in the universe today, including that for the highest density. But there's one cosmic event that surpasses even the black holes' superlatives. About 13.8 billion years ago, all of the matter and energy we see around us today burst into being in that most inconceivable of phenomena: the Big Bang. The earliest meaningful time we can talk about in science today is one ten million trillion trillion trillionth, or 10^{-43}, of a second after the point at which time began. At this earliest moment, gravity split away from the other three fundamental forces of nature. All of the mass and energy we observe in the universe today was corralled within a volume of space just 1.6×10^{-35} of a metre across – far, far smaller than an individual proton or neutron. Little wonder that its density was higher than anything that ever has been, or ever will be, subsequently matched – a staggeringly high 10^{90} kilograms per cubic centimetre.

CHAPTER 21

Black

WHEN I WAS young I went down Blue John Cavern in Castleton, Derbyshire, a public show cave near where I lived. The guide takes you to a depth of more than 200 feet into a cathedral-like space, amid stunning rock formations and stalactites – and then turns out the lights. With not the faintest ray penetrating from the surface and no artificial source of illumination, you're immersed in total, absolute blackness. But this is blackness due to the absence of light. A surface or material that's completely black may be bathed in light yet look as utterly dark as the deepest cave.

Just as white light is a mixture of all the colours of the rainbow, black is the absence of any of those colours. Like white, it's sometimes considered to be a colour in its own right. But because it doesn't appear anywhere on the visible spectrum, from red to violet, it's technically a shade. Whatever we choose to call it, it provides the best contrast with white or any light colour and so has been used to make written communication stand out since the dawn of human history.

Black has both practical and artistic applications. It's also come to represent certain qualities, such as solemnity and authority, and to symbolise death and mourning. Its importance has encouraged the search for ways to make darker, more durable black substances and, in recent times, for the ultimate blackest of blacks.

Even before written language, cave painters at Lascaux and other Palaeolithic sites turned to black pigments for making their striking images of large animals and hunting scenes. To begin with, they used the charcoal from fires but later burned bones or ground a powder of natural manganese oxide as a source of drawing material. Animal fat and resin served as a binder.

The earliest black inks were made from lampblack, ground-up dark minerals such as graphite, and natural pigments from plants and animals suspended in water. Over time, different cultures discovered how to make ink that was thicker, more permanent and less prone to fading. In the same way, improved black dyes were found for clothing and other fabrics. Often, complex chemical and physical processes were

An Athenian black-figure pottery amphora, from the sixth century BCE, depicting Theseus slaying the Minotaur.

involved, which were discovered over time by trial and error. The black in Greek black-figure pottery, for instance, was produced by a sensitive three-phase firing process carried out at different temperatures.

The blacker the better for some purposes. The interior walls of optical instruments, such as telescopes, microscopes and cameras, must reflect or transmit as little light as possible so as not to interfere with the final image. With the development of specialist and highly sensitive optical instruments for scientific and military purposes has come an acute need for extremely black surfaces. Space agencies, such as NASA, and defence departments are among those who've funded research into blacker blacks, for example, to prevent stray light from entering telescopes and improve the performance of infrared cameras both on Earth and in space.

In 2002, researchers at the National Physical Laboratory (NPL), in Teddington, London, announced that they'd developed a new surface treatment called super-black. More than two decades earlier it had been discovered that chemically etching a nickel-phosphorus alloy could produce a more intense black. But not until the breakthrough at NPL, which came after hundreds of tests, had anyone managed to come up with a nickel-phosphorus formulation and etching process that was better at absorbing light than the blackest paint.[1]

Super-black reflects ten to twenty times less light than the most effective black paint that's used to reduce reflections in instruments. To produce it, the object to be blackened is immersed for several hours in a solution of nickel sulphate and sodium hypophosphite. This results in a nickel and phosphorus coating, the surface of which is then etched with nitric acid to create the super-black surface. The percentage of phosphorus in the coating is crucial to how the surface behaves after etching. More than 8 per cent and the surface has a spiky texture, but if the phosphorus content is about 6 per cent, it's covered in tiny craters. These microscopic pits are much better at not returning any light that falls upon them.

When light strikes at right-angles to a super-black surface, less then 0.35 per cent is reflected back – seven times less than from black paint. At greater angles the advantage increases. At an angle of 45 degrees, super-black reflects twenty-five times less light than black paint. It's especially useful for coating parts of instruments that must be kept cold, such as infrared space telescopes, because unlike ordinary paint it doesn't crack at low temperatures.

In 2014, a substance that's even blacker than super-black was announced. It's known as Vantablack, VANTA being an acronym of 'Vertically Aligned NanoTube Arrays'. Carbon nanotubes were discovered by the Japanese physicist Sumio Iijima during experiments in 1991. Each consists of a hollow tube just a few, or a few tens of, nanometres (billionths of a metre) in diameter, with a cylindrical wall made of a mesh-like arrangement of carbon molecules.

Early on it was found that carbon nanotubes have some extraordinary physical properties. They're exceptionally good conductors of both electricity and heat, and have a very high tensile strength. Then came the discovery, by British engineer Ben Jensen, who founded the company Surrey NanoSystems, that large numbers of nanotubes, when aligned, like trees in a forest, are incredibly effective at absorbing light.[2]

Jensen's original Vantablack coating was made by a process known as chemical vapour deposition (CVD), which took place in a reactor vessel at 400 °C and in a near-vacuum. Millions upon millions of carbon nanotubes, each about 10 nanometres in diameter and 30 micrometres long, were formed as a surface layer. The coating absorbed up to 99.965 per cent of visible light falling on it, whatever the viewing angle. The result is astonishing. An object coated in Vantablack shows no relief – no wrinkles, indentations or features whatsoever.

Controversy erupted when British-born Indian sculptor and installation artist Anish Kapoor obtained exclusive rights from Surrey NanoSystems to use Vantablack as an art material.[3] Many of his peers were outraged that Kapoor would try to deny access, for creative purposes, to the purest of blacks. British artist Stuart Semple protested by developing a pigment he called 'Pinkest Pink' and making it available on his website with the legal rider:

> By adding this product to your cart you confirm that you are not Anish Kapoor, you are in no way affiliated to Anish Kapoor, you are not purchasing this item on behalf of Anish Kapoor or an associate of Anish Kapoor. To the best of your knowledge, information and belief this paint will not make its way into the hands of Anish Kapoor.

In 2019, BMW partnered with Surrey NanoSystems to create the VBX6, a version of the company's X6 series coupé spray-painted with a newer form of Vantablack that's easier to apply than the original. Aside from its windows, wheels and lights, the car looks from a distance to be almost two-dimensional, like a silhouette.[4]

In 2019 also came the announcement of a material that's even blacker than Vantablack. It too uses vertically aligned carbon nanotubes but it involves a different manufacturing process.[5] Engineers Brian Wardle and Kehang Cui at the Massachusetts Institute of Technology didn't set out trying to make a darker black. Rather, they were experimenting with ways to grow carbon nanotubes on electrically conducting materials such as aluminium, to improve their electrical and thermal properties. But in trying to grow nanotubes on aluminium, they ran into a problem: a layer of oxide always forms on the metal when it's exposed to air. This layer insulates rather than conducts electricity, so it had to be removed.

The MIT group had already been using common materials, such as table salt (sodium chloride) and baking soda, to grow carbon nanotubes, and it's well known that seawater dissolves oxide on the surface of aluminium before slowly eating away at the metal itself. Adopting this approach, the researchers soaked aluminium foil in saltwater to remove the oxide layer then put the foil in an oven to grow nanotubes by CVD. They found what they expected – that the nanotubes boosted the electrical and thermal conductivity of the coated metal. But when they measured the amount of light reflected from the material they were astonished by how little there was.

The MIT nanotube/aluminium combination absorbed at least 99.995 per cent of all incoming light, with equal effectiveness from every direction. It was blacker even than Vantablack. Quite why it's so dark is an open question but it is evidently due to the combined effect of the aluminium, which is somewhat blackened during the etching process, and the carbon nanotubes.

As with Vantablack, the MIT coating is finding uses in both technology and art. Diemut Strebe, an artist-in-residence at the MIT Center for Art, Science and Technology, has collaborated with Wardle and his group to coat a large yellow diamond, worth around $2 million, with the new ultra-black material. The contrast between the gem in its normal state, with brilliant, sharply defined facets, and its alternative, utterly blackened form, featureless and impenetrably dark, is remarkable.

NASA senior astrophysicist John Mather sees a more practical use for the ultimate black coating, in the search for exoplanets. We receive so little light from worlds going around other stars that in order to make them visible it's imperative to cut out any unwanted glare. The material developed at MIT would be ideal, Mather has pointed out, on the surface of a large star shade to shield a planet-hunting space telescope from stray light.

We think of space itself as being black but, of course, in reality, it's filled with the light given off by countless stars. Even the intergalactic void, where stars are few and far between, is crisscrossed by vast numbers of photons – particles of light – coming from all parts of the universe. Only one type of object is darker than anything we've yet mentioned. By definition, a black hole is a region of space from which nothing, not even light, can escape. Black holes are perfect absorbers of light, the darkest of the dark.

CHAPTER 22

Reflect on This

THE FIRST MIRRORS were simply pools of still, dark water. They would have been almost the only way to see yourself in prehistoric times because for a surface to give a true reflection it must be very flat and smoother than the wavelength of light (between about 40 and 70 millionths of a centimetre).

The only mirrors found in nature, apart from water, are the flat surfaces of some crystals and forms of natural glass. The latter include obsidian, a dark, naturally occurring glass formed when lava, erupted from a volcano, cools very quickly. Among the earliest known manufactured mirrors are ones made from obsidian discovered in Anatolia (modern-day Turkey) and dating to about 6000 BCE. By 4000 BCE, craftspeople in Mesopotamia were fashioning mirrors from burnished copper; a thousand years later, similar artefacts were being made in Egypt.

By the Bronze Age, many ancient cultures were producing mirrors from a variety of metals, including silver – the most reflective natural metal (and element) of all. Polished silver reflects about 95 per cent of the visible light that falls upon it. In time, that would make it desirable for use in making not just vanity mirrors but also the reflecting surfaces of scientific instruments such as telescopes and microscopes.

The earliest known mirrors made from manufactured glass date from the third century CE but appear to have been used only as jewellery or amulets. They consist of curved metal surfaces with glass coatings and are very small. Making practical glass mirrors is no easy task. Once you have molten glass, how can you form it into thin, flat panes that are both clear and untinted? Even trickier, how can you then apply molten metal to a sheet of glass without causing it to crack and break? It took until the twelfth century for workable techniques of plane glassblowing and silvering to be developed. By the late Middle Ages, Venice became the glass manufacturing and mirror making centre of the world.

The first telescope to use a mirror to gather and focus light – a reflecting telescope – was invented by Isaac Newton in 1668. But it didn't perform all that well. First, it was very small, with a diameter of only 3.3 centimetres (1.3 inches). Second, it had a circular curvature, which meant that parallel rays of light entering the telescope and striking different parts of the mirror weren't all focused at the same point. To bring all the light rays to a single focal point the mirror must be parabolic in shape.

Newton's mirror was made of speculum metal. This is an alloy of about two-thirds copper and one-third tin, which can be polished to make it highly reflective. It was the only material available at the time that could be made into a reasonably accurate curved reflecting surface. Glass could be ground into a curved shape but the reflecting surface, in the form of a layer of metal, had to go at the back and was flat.

All the reflecting telescopes that followed Newton's, until the middle of the nineteenth century, used speculum mirrors and some grew very large indeed. In 1789, William Herschel built his '40-foot telescope' using a mirror with a diameter of 126 centimetres (49.6 inches). Largest of all was the 'Leviathan of Parsonstown', built in 1845 by William Parsons, 3rd Earl of Rosse, on his estate, Birr Castle, at Parsonstown (now Birr) in Ireland.[1] Its mirror was 1.83 metres (72 inches) in diameter and the largest in the world until the completion of the 2.5-metre (100-inch) Hooker Telescope in California in 1917. The Leviathan's tube and mirror box were 16.5 metres long and, with the mirror in place, weighed 12 tons.

Parsons pushed the technology of casting, grinding and polishing large telescopes from speculum metal to its limit at the time. He built steam-powered grinding machines to achieve an accurate parabolic shape. But speculum has some serious disadvantages. It's difficult to cast and shape, reflects only two-thirds of the light that falls on it, and tarnishes quickly in open air. This last fact meant that a spare mirror had to be fabricated for each telescope to replace the other while it was being polished and reconfigured.

A major breakthrough came in 1856 when the German physicist and astronomer Carl von Steinheil developed a way of depositing an ultra-thin layer of silver on the front surface of a piece of glass. Almost overnight speculum mirrors became obsolete because now the ease and accuracy with which glass could be ground to a precise curvature could be combined with the high reflectivity of silver. The only drawback of silver is that, once exposed to air, it immediately begins to tarnish by combining with oxygen and forming a slender coat of silver oxide. Polishing and re-silvering has to be done regularly to maintain the mirror's performance.

In 1930, American physicist and astronomer John Strong invented a process to lay down aluminium on glass, allowing much more reflective mirrors for telescopes. Once a mirror has been ground and polished to its final, accurate parabolic shape, it's put in a vacuum chamber with electrically heated coils that cause aluminium to evaporate. In a vacuum, the hot aluminium atoms travel straight to the surface of the mirror where they cool and stick. Although, like silver, aluminium oxidises in air, the thin aluminium oxide is transparent and so the underlying metal can continue to reflect light efficiently.

Most large telescopes in the world today use mirrors coated in aluminium – but not all. Some telescopes launched into space, such as the Kepler Space Telescope, use silver because in the vacuum beyond Earth's atmosphere oxidisation isn't an issue. And, in Kepler's case, it was vital that every bit of reflectivity be squeezed out of the mirror because the instrument had to be able to detect the faintest trickles of light from far-off exoplanets. Among big ground-based telescopes, the ones at the Gemini Observatory are unusual in having been silvered rather than aluminised. To get around the problem of tarnishing, an annual multi-layer protective coating process is employed.

In 1949, the largest telescope in the world was the newly completed 5.1-metre (200-inch) reflector on Palomar Mountain, near San Diego, California. But today's giant instruments are in the 8- to 11-metre range, with even mightier ones on the way. These new scopes use designs and technologies that were unavailable until the last quarter of the twentieth century, including the use of segmented mirrors, in which each facet is independently controlled by computers to ensure correct focusing of the final image.

The European Extremely Large Telescope (ELT), currently being built on top of a mountain in the Atacama Desert in northern Chile, will consist of a segmented mirror with an effective diameter of 39.3 metres. It will be able to gather 100 million times more light than the human eye and, even though it has to peer through the atmosphere, will provide images that are sixteen times sharper than those from the Hubble Space Telescope.

The largest telescope off the planet is the James Webb Space Telescope, launched in December 2021, which has a 6.5-metre-diameter segmented mirror with a collecting area six times that of Hubble.[2] Webb's mirror has a gold-plated beryllium surface, which is particularly effective at reflecting in the infrared part of the spectrum where the telescope is designed to operate.

When scientists talk about how an object or surface reflects sunlight they often refer to a quantity called albedo. This is measured on a scale that goes from 0, corresponding to complete blackness, to 1, if a body is bright white and reflects all the radiation that falls on it. Charcoal, one of the blackest natural substances, has an albedo of 0.04, whereas fresh snow scores 0.9.

The average albedo of Earth is about 0.3. This is much higher than the value for areas of ocean alone because of the contribution of clouds and the polar ice caps. Venus has an albedo of 0.76 because its surface is entirely hidden beneath a dense atmosphere and 100 per cent permanent cloud cover.

Enceladus, the most reflective moon in the Solar System, as imaged by the *Cassini* probe

The brightest object of all in the Solar System is Saturn's moon Enceladus, almost the entire surface of which is covered in water-ice. The general lack of craters on its smooth plains suggests that these regions are less than a few hundred million years old – young by geological standards – so that some process, such as water volcanism, is renewing the surface. The fresh, clean ice and snow gives Enceladus an albedo of 0.81. A consequence of reflecting so much sunlight is that among all Saturn's moons, it has the lowest temperature: at noon, a chilly −198 °C.

CHAPTER 23

Slip Sliding

'A BALL IS moving on a horizontal frictionless plane...'
'A child and sledge slide down a frictionless hill...' Anyone who's studied physics in their final years at high school will have come across problems like these in which we pretend, to make things simpler, that there's no friction.

In the real world, friction is always present – thankfully. Without it, cars wouldn't be able to move, we couldn't take a single step, and pretty much nothing would stop until it crashed into something else! But the amount of friction varies a lot from one substance to another, and some materials slip or slide very easily when in contact with another surface.

It's a well-known fact that when you push a heavy object, like a sofa, it's a lot easier to keep it moving than it is to get it started moving in the first place. The Greek philosopher and statesman Themistius, in 350 CE, put it this way: 'it is easier to further the motion of a moving body than to move a body at rest.' A quantity called the coefficient of friction measures how easily two surfaces or materials will slide against each other. It comes in two main versions: static and kinetic. The coefficient of static friction indicates how hard it is to start movement between two surfaces, while the coefficient of kinetic friction is a factor in the effort needed to keep the motion going.

We use rubber for the soles of shoes and the tyres of road vehicles because it's flexible and grips well. So, not surprisingly the coefficient of friction between rubber and, say, concrete is pretty high: 1.0 in the static case, and 0.6 to 0.85 when rubber is moving against concrete. Compare that with the coefficient of friction for steel on ice, which is 0.03.

From an early age we learn, often the hard way, that ice is slippery. It's such a well-known fact that it's surprising to find that scientists still debate the cause of its slipperiness. An old theory, going back to the nineteenth century, is that the weight of something pressing down on ice melts a thin layer of it so that the object can glide along through the water molecules that it's freed up. But that doesn't explain, for instance, why it's possible to skate on ice that's very cold – much too cold for water to form by so-called 'pressure melting'. You'd also need a ridiculously high pressure, like that of 'an elephant in stilettos', as one scientist put it, for this mechanism to work.

Other ideas have been put forward over the years to account for this most basic of observations: that things slide easily on ice. It's a question central to studies in how glaciers move, automotive safety and winter sports performance. But only in 2018 was a study presented that may, once and for all, have cleared up the mystery. Physicists and brothers Mischa and Daniel Bonn, and their colleagues, published a paper in the *Journal of Chemical Physics* describing how the surface of ice is slippery not because it consists of a layer of water but because it has loose water molecules.[1] They compared the ice surface to a dance floor covered with marbles so that slipping on ice is like rolling on round, mobile molecules.

For the most part, ice has an orderly crystalline structure in which each molecule is firmly attached to three neighbouring ones. But at the surface, the molecules have fewer attachments and so are only weakly bound to the crystal. This allows the surface molecules to tumble, and to repeatedly connect and disconnect with the lattice as they go.

Ice isn't unique in being slick because of loose surface molecules. But it is unusual in that we usually experience it in a temperature range where it can exist as a solid, liquid or gas. In places where it gets very cold, ice starts to behave differently. Below about −40 °C, the molecules at the surface have less energy to move around and tend to become attached to the crystal with the result that the ice loses its slipperiness. This new research found that ice offers the least resistance at around −7 °C. That came as no surprise to those who maintain indoor skating rinks, especially those used for speed-skating, who had already discovered this optimum temperature by trial and error.

Living things have evolved various kinds of non-stick substances for their own advantage. Goblet cells in the lining of our gastrointestinal tract and lungs secrete mucous that serves to lubricate surfaces for ease of movement. In some species, the amount of slime an animal can produce, often as a means of defence, is spectacular. The hagfish – a primitive, eel-like fish – is unequalled in this regard.[2] When stressed, it can release goo from special glands, which, in less than half a second, expands over 10,000 times in volume. In one memorable incident, in 2017, a truck carrying 3,400 kilograms of hagfish in thirteen containers braked sharply on an Oregon highway, spilling its load. Moments later the road and a nearby car were dripping in thick, white slime, while several other vehicles slid and collided on the glistening gloop.

Not to be outdone, plants too have their reasons to be slippery. The carnivorous pitcher plant lures unwary insects onto its specialised leaf – called a pitfall trap – with nectar and attractive pigments. But having settled on the lip of the tempting foliage the victim quickly slides down its steep walls into a pool of liquid at the bottom.[3] Escape is impossible and the hapless prisoner is then dissolved in digestive juices to feed its vegetative captor.

Members of the genus *Nepenthes*, or tropical pitcher plant, attracted the attention of a group of researchers at Harvard University. The leaves of *Nepenthes* are slick because of a thin lubricating film that lines their inner surface. The film forms when water or nectar becomes locked into microscopic scales on the leaf, creating a continuous layer of lubrication. When the oils on the feet of an insect make contact with the film the friction is very low, making it almost impossible for the creature to maintain a grip when trying to climb out.

Mimicking *Nepenthes*, the Harvard scientists developed an 'omniphobic' material which they described as a 'slippery liquid-infused porous surface, or SLIPS'. It consists of a sponge-like network of nanofibres coated in a lubricating film that repels a wide range of liquids.

Some substances are exceptional in more ways than one. BAM (short for Boron Aluminium Magnesium) was discovered accidentally in 1999 by scientists at the US Department of Energy. They were trying to create a material that would generate electricity when heated. Although BAM doesn't do that, it turns out to be extremely hard – almost as hard as diamond – and incredibly slippery. In fact, it has the lowest coefficient of friction of any known substance – less than half that of the previous record-holder, Teflon. Known chemically as polytetrafluoroethylene, Teflon was discovered serendipitously by a researcher at DuPont in 1938 and has a coefficient of friction of 0.04. BAM maintains an unparalleled 0.02 and can be applied as a microscopically thin coating to lots of different surfaces to help them slide easily over other surfaces. The drastic reduction in friction saves energy and extends the life of moving components.

Friction is such a familiar part of the everyday world that it's hard to imagine it not being there. It acts on anything that moves on land, in water and through the air. We assume that all objects will eventually come to rest unless some force continually pushes or pulls them along. But in the vacuum of space there's no friction and so spacecraft, moons, asteroids and everything else continues on its way not slowed down or hindered in any way by air resistance.

Friction with the air is why feathers fall more slowly than stones. Take away the air, as on the Moon, and we see how everything falls at the same rate under gravity. This was beautifully demonstrated during the Apollo 15 mission when commander David Scott, standing on the Moon, dropped a feather and a hammer together and the two hit the ground at the same instant.

Achieving zero friction in a substance may sound impossible. Even BAM, the slipperiest of slippery materials, has a coefficient of friction that's more than nothing. But a state, known as superfluidity, does exist in which the atoms move around and slide past one another without losing any energy. Only two superfluids made of ordinary matter are known and both are forms of helium.

In 1937, physicists Pyotr Kapitsa, John F. Allen and Don Misener discovered that helium-4 enters a superfluid phase at very low temperatures. Later on it was shown that if little whirlpools or vortices are started in helium-4 in this state, they'll continue to spin around forever. The superfluidity of helium-3 was discovered more recently and was the subject of the 1996 Nobel Prize in Physics.

CHAPTER 24

Slo-mo Flow

IN THE 1941 movie *Gone with the Wind*, Scarlett O'Hara chides Prissy for being as 'slow as molasses in January'. The expression seems to have originated in the US sometime in the middle of the nineteenth century. And it's true that molasses, known commonly in the UK as treacle or black treacle, is a thick, syrupy liquid that drips slowly off a spoon, especially when cold.

But unless you've experienced molasses *en masse*, as it were, even in the depths of winter, you might want to avoid repeating Miss O'Hara's phrase. Picture this: it's 12:30 pm, January 15, 1919, the temperature is 6 °C, and the place is Boston, Massachusetts, on a low-lying section of Commercial Street between Copps Hill and North End Park. A sticky situation is about to unfold: a 58-foot tank that stood behind the Boston and Worcester freight terminal has just burst open, releasing 2.5 million gallons of molasses. A small tin of treacle is one thing. A 10-metre-tall tidal wave of the stuff advancing towards you at over 50 kilometres per hour is an altogether more impressive affair. In the Great Molasses Flood, 21 people died, 150 were injured and the steel supports of the nearby elevated train structure were ripped away. More than memories of the dreadful event lingered. For decades afterwards, locals claimed that the area still smelled of molasses on hot summer days.[1]

Molasses, golden syrup and honey are among the liquid foods that have a high viscosity, especially at low temperatures. Viscosity is the resistance of a fluid – a liquid or gas – to flow. It's a measure of the internal friction acting between the molecules of a substance as it moves. The word comes from the Latin *viscum* for 'mistletoe' and derives from the name of the viscous glue that can be made from mistletoe berries.

In the international system (SI) of units, viscosity is measured in pascal-seconds (Pa.s), the pascal being a unit of pressure, millipascal-seconds (mPa.s) or micropascal-seconds (μPa.s). The higher the viscosity, the more sluggishly a substance moves or the more difficult it is for something to move through that substance. In general, viscosity decreases with rising temperature or, as illustrated by the Great Molasses Flood, with increasing pressure.

Water has a viscosity of 1.0 mPa.s. As you might expect, gases have very low viscosities, measured in micropascal-seconds. At room temperature, olive oil comes in at 56 mPa.s, honey (depending on its consistency) at 2,000–10,000 mPa.s, molasses 5,000–10,000 mPa.s, and ketchup 5,000–20,000 mPa.s. A collective sigh of relief came when ketchup was finally made available in squeezable plastic containers instead of the glass bottles that delivered, after much shaking, a dollop of the red stuff of unpredictable size and trajectory.

Moving up the viscosity league we come to seriously slow-moving liquids. You may not think peanut butter is a liquid but it is – as you'll quickly be reminded at airport security if you have a jar of it in your carry-on luggage. This favourite of American spreadables – viscosity in the range 100,000 to 1 million mPa.s – will eventually display its liquid credentials if you tip a jar of it upside-down and wait long enough. Unfortunately, airport scanners can't tell the difference between a harmless liquid and one that might be explosive. And, as it happens, the atomic constituents of peanut butter are pretty much the same as those of nitro-glycerine – mostly carbon, hydrogen, nitrogen and oxygen.

In Chapter 2, we saw how pitch is at the heart of the longest-running experiments in science, at universities in Australia and Wales. In the University of Queensland pitch-drop experiment, nine drops have fallen since 1927, allowing researchers to estimate the pitch's viscosity at 23 billion times that of water.

It's sometimes said that glass is a kind of liquid – a 'supercooled' liquid – and that it flows very slowly over time. Evidence for this, it's claimed, is that old windows, such as those in mediaeval cathedrals, are sometimes found to be thicker at the base than the top. In fact, this is wrong.[2] Any differences in the width or other non-uniformities of old glass were present on the day it was made. It's just that the manufacturing processes used many hundreds of years ago resulted in panes with imperfect surfaces and uneven thickness. Glass isn't a liquid at all but what's known as an amorphous solid. Its rate of flow would be impossible to measure – less than 1 nanometre in a billion years.

Other solids also flow under the right circumstances of temperature and pressure. Glaciers are often described as 'rivers of ice', moving downhill and deforming in response to gravity. Glacier ice effectively flows like a liquid with a very high viscosity – about 1,000 Pa.s.

Even solid rock will flow given enough time and subjected to enough stress. This type of movement is known as rock creep and it means, strangely enough, that substances as seem-ingly hard and solid as granite can be said to have a viscosity.

CHAPTER 25

Poison Most Deadly

IN THE ANNALS of crime, fictional and real, certain poisons crop up again and again. Arsenic has been a favourite of murderers since Roman times when it was used to dispose of political rivals and even emperors. White arsenic – arsenic oxide – obtained as a by-product of copper and lead refining, is both tasteless and water-soluble, making it easy to slip unnoticed into drinks. The victim would soon experience stomach pains, followed by severe vomiting and diarrhoea, as if suffering from food poisoning. Circulatory collapse and death would ensue within hours.

In Renaissance times, poisoning became a lucrative profession. Clients would be given a price and sign a contract, after which the victim's fate would be sealed. Some of the Borgia family, including Pope Alexander IV, his son Cesare, and Cesare's half-sister, Lucrezia, were notorious for this form of assassination. In seventeenth-century Italy, Giulia Toffana made a living by supplying *Aqua Toffana*, an arsenic-laced cosmetic, along with instructions on how to apply it, to women wishing to dispose of their abusive husbands. In France, a similar concoction nicknamed *poudre de succession* ('inheritance powder') became a popular means by which women married to wealthy men could suddenly find themselves eligible widows.

The incidence of murder by arsenic poisoning declined in the nineteenth century following the discovery of a way to detect it during post-mortem examination. Yet arsenic products continued to be widely used in quack cures and cosmetics. Victorian women applied arsenic powder to whiten their faces, while 'arsenic complexion wafers' promised to remove freckles, pimples and other skin imperfections. Fowler's solution, containing potassium arsenite, was favoured by prostitutes of the time because it gave them rosy cheeks – an effect caused by damage to the capillaries.

Atropine, or 'belladonna', was the go-to poison for other murderers in mediaeval Europe, the juice of just a few berries of the deadly nightshade being enough to achieve the dastardly effect. In small doses, it's been used as a hallucinogenic as far back as the ancient Greeks. Larger amounts produce symptoms that, conveniently for the killer, mimic those of various fevers that were common in those days.

Of the traditional, well-known poisons, cyanide is the fastest acting. It kills within minutes, making it ideal for use in suicide pills of the type carried by spies and Nazi German officials in World War II in the event of being captured. It can be distilled from the kernels of almond nuts and is also found in the leaves of some laurel bushes. In 1982, Tylenol capsules laced with potassium cyanide led to seven deaths in the Chicago area – no one was ever charged or convicted of the crimes. Subsequently, hundreds of copycat attacks took place around the United States, which led to reforms in the way over-the-counter pharmaceuticals were packaged.

In Agatha Christie's novel *The Pale Horse*, a spate of mysterious deaths is eventually tracked down to the administration of thallium – a poison that was little-known before publication of the book in 1961. In fact the element itself had been discovered only about a century earlier. Thallium sulphate became an agent-of-choice for assassins: perfect for the job because it was tasteless in solution and took several days to produce symptoms, which mimicked those of other illnesses. The Russian KGB and Saddam Hussein's secret police were especially fond of it.

Although there's no evidence that Christie's novel inspired any murders by thallium, it certainly saved one or two lives. In 1977, the *British Journal of Hospital Medicine* reported the case of a nineteen-month-old girl admitted to Hammersmith Hospital with a mysterious ailment that was worsening by the day.[1] Doctors had no clue what was wrong with her. But then a nurse at the hospital, Marsha Maitland, came across the girl and recognised her symptoms. They were the same as those described in *The Pale Horse*, which, by good fortune, Maitland had recently read. A urine sample was sent to Scotland Yard, which confirmed the presence of thallium. It turned out the girl had accidentally swallowed some insecticide containing the deadly element. Doctors were then able to treat her appropriately and, in time, she made a complete recovery.

A common way to measure the strength of a poison is the 'mean lethal dose', or LD_{50}. This is the amount of a substance in milligrams per kilogram of body weight needed to kill 50 per cent of a given population. Any substance, even water, will kill you if you ingest enough of it in one go. Table sugar, for instance, has an LD_{50} of 29.7, which means that half of a group of people each crazy enough to eat about 30 grams of it for every kilogram they weighed would probably suffer a not-so-sweet demise as a result. Sodium cyanide, on the other hand, has an LD_{50} of 0.0064. So, if you weighed 70 kilograms, a dose of just a third of a gram – a small pinch – would have a 50:50 chance of killing you.

To enter the big league of poisonous substances, though, we have to move away from inorganically produced chemicals and look instead at *toxins*: poisons manufactured by living things. Take the poison arrow frog – or, rather, don't take it. Don't even think of touching it. Oozing from its skin is batrachotoxin, a chemical that would prove fatal even in the tiny amount you'd ingest from licking a finger after handling the animal.

Interestingly, captive-born frogs of the various poison arrow species are completely harmless. That's because the toxin isn't produced by the frogs themselves but comes, instead, from the type of beetles they eat in the forests of Central and South America. By chance, American ornithologist Jack Dumbacher found that the pitohui, a songbird living on the other side of the world, in Papua New Guinea, also carries batrachotoxin in its plumage – and for exactly the same reason as the frogs.[2] He made the discovery after being scratched on the hand by one of the birds. Instinctively, he put his hand to his mouth, which started to go numb. It seems the birds have evolved to use the poison as a chemical defence against parasites or predators.

Another creature to avoid touching is the pufferfish. It's covered in spines that drip tetrodotoxin, with an impressive LD_{50} of 0.0000082, making it more than a thousand times as toxic as cyanide. An almost invisibly small amount of it would be enough to kill a person within an hour. If you plan a meal of pufferfish, or fugu as it's called in Japanese, then pick your eatery with care. In Japan, only chefs who've qualified after at least three years of training are legally allowed to prepare the dish and, happily, restaurant mishaps are extremely rare. Most victims are anglers who rashly try to fillet and consume the fish at home. Tetrodotoxin poisoning has been described as 'rapid and violent', beginning with numbness around the mouth and progressing to paralysis and death. There's no antidote and the unfortunate diner remains conscious to the end.

One of the many types of pufferfish – a specimen of
Diodon nicthemerus, also known as a porcupinefish.

Some plants and fungi, too, are highly poisonous. Among
them, the appropriately named death cap (*Amanita phal-
loides*) is the single species responsible for most fatalities
worldwide. Eating just half a death cap mushroom can
kill due to the amatoxins, with an LD_{50} of 0.0007, they
contain. Nor would cooking the fungus save you. Unlike
many ingested poisons, amatoxins can't be destroyed by heat
without rendering the mushrooms inedible. For up to half
a day after swallowing a death cap you'd seem fine. Then
gastrointestinal upset would start – vomiting and non-stop
watery diarrhoea – and go on for twenty-four hours. After
this, the amatoxins would go to work on your liver, possibly
damaging this vital organ to the point where you'd need a
liver transplant to survive.

Unsurprisingly, poisonous substances, designed to kill
either on a large scale or with specific deadly accuracy, have

found their way into military arsenals. Their first use came during World War I when Germany released chlorine gas from thousands of cylinders across a 6-kilometre-wide front at Ypres on April 22, 1915. In many cases, the various gases used in the 1914–18 conflict, including phosgene and mustard gas, didn't kill but instead inflicted horrible injuries or disfigurements on their victims.

The Second World War saw the emergence of nerve agents, which target the mechanism that enables nerves to transfer messages to the body's organs. Among the first group of such chemicals, called the G-series because German scientists were the first to develop them, was Sarin – lethal within minutes, due to respiratory paralysis, even at very low concentrations. Production and stockpiling of it was outlawed by international agreement in 1997, though it has been used since, most infamously in 2013 in an attack near Aleppo in Syria, which killed twenty-eight and injured more than a hundred others. Several assassinations in recent times have involved Novichok nerve agents, developed in the Soviet Union between 1971 and 1993.

At the extreme high end of toxicity are certain incredibly potent substances produced by bacteria. Scientists debate the exact ranking of some of the nastier strains, such as diphtheria toxin, Shiga toxin and tetanospasmin. But everyone seems to agree on the deadliest poison of all – botulinum toxin, a chemical produced by the bacterium *Clostridium botulinum*. It boasts the smallest LD_{50} of any substance known, just 0.000000001, so that an intravenous dose of a mere ten-millionth of a gram would be enough to kill an average-sized person.

Botulinum toxin, of which several types are now known, is responsible for botulism, first identified as a cause of food poisoning in a small German village in 1793. The toxin paralyses muscles by preventing the release of the neurotransmitter (a signalling molecule) acetylcholine. This same paralysing property explains its use in Botox. Targeted injections of the toxin, in tiny quantities, relax certain muscles and give the appearance of smoother skin, though only temporarily and with the risk of side-effects.

Some of the most toxic chemicals on Earth can be used to medical advantage. Aside from its well-known cosmetic use, Botox has been applied to clinical conditions, such as treating strabismus, or squint, by paralysing the muscles that cause the eyes to point in different directions. The venom of the lethal Brazilian pit viper, *Bothrops jararaca*, contains blood-pressure-reducing molecules that have led to pioneering treatments for hypertension.

Paracelsus summed it up well 500 years ago: 'All things are poison, and nothing is without poison: the dosage alone makes it so a thing is not poison.' We live amid countless potentially dangerous substances – it's just the amount we ingest that can make them deadly.

CHAPTER 26

You're So Sweet

IN THE TOWN where I went to secondary school – New Mills in Derbyshire – many of the older students had part-time jobs at a local company called Swizzels. It had its roots in the 1920s in a market stall in Hackney before expanding to a manufacturing site in East London. To escape the Blitz, the company then moved north to a disused factory in New Mills, where it remains to this day. Swizzels makes sweets: Refreshers, Parma Violets, Drumstick lollies and, best known of all, Love Hearts.

Sugars have always been part of our diet. They occur naturally in fruits and some vegetables in the form of fructose, glucose and sucrose. Less obviously, milk contains another type of sugar – lactose. We've been consuming sugars for as long as we've been human and there's plenty of evidence that we'll go out of our way to get them. Honey, which is rich in fructose and glucose, has been a delicacy worth risking a few bee stings to obtain since our Stone Age, cave-dwelling days. Cave paintings in the Cuevas de la Araña in Spain show humans foraging for honey at least 8,000 years ago.

By 2000 BCE, the Egyptians were combining honey with fruit and nuts to make an early form of confectionery. Between the sixth and fourth centuries BCE, word reached the Persians then the Greeks that from the Indian subcontinent came 'reeds that produce honey without bees'. Knowledge of how to grow this wonderful crop – sugar cane – then spread westward from its literal roots in India and Southeast Asia.

Sweetness is one of five distinct types of taste that we can experience, the others being sourness, bitterness, saltiness, and savouriness or umami. These different modalities are detected by several thousand taste buds, each containing up to 100 receptor cells, located on the front and back of the tongue. Others are found located on the roof, sides and back of the mouth and in the throat.

Some animals can taste things that we can't. What does starch taste like? Ask a rat. But don't try to tempt a cat with sweet foods because sweetness isn't in their taste repertoire. Bottlenose dolphins are even more taste blind: they can perceive only saltiness.

Humans have a very obvious 'sweet tooth'. Equally obvious is the fact that too much sugary stuff is bad for us, in all sorts of ways. So why are we so attracted to it if it can be harmful? Evolutionary forces generally encourage us to like things that are beneficial – if not to us personally, then to our species as a whole. Many sweet substances in nature, if consumed, are ultimately a source of glucose in our bodies, and glucose is an energy powerhouse, especially as a fuel for the brain.

Sugars that, once eaten, can be broken down into glucose occur in all fruits. The lactose in milk also gets converted into glucose, which is one of the reasons babies are motivated to gulp it down almost from the moment of birth. So, our sweet taste, the theory goes, evolved as a way to detect sources of glucose. But our liking for sugars is only part of the story. We tend to avoid foods that are bitter because, evolution has taught us, these are often indigestible or downright bad for us because they're poisonous or rotten. While it's true that some bitter substances can be beneficial, for instance in the case of medicinal plants, in general there's a correlation between the toxicity of a compound and its bitterness. In the overall scheme, then, what matters is that we prefer sweetness to bitterness because that improves our survival chances. This innate bias spills over even into character descriptions: a person who's sweet is far more pleasant to be around than one who's bitter!

Nowadays, table sugar – pure sucrose – is almost too plentiful for our own good. But it wasn't always that way. A method of producing sugar crystals was known in India by about 500 BCE. In the local language, these crystals were called *khanda*, from which comes our word *candy*. Sugar came to Europe in the Middle Ages via the Arab world but at first was regarded solely as a medicine, supplied by apothecaries and used by physicians. It was also extremely expensive and therefore available only to the wealthy. In 1390, the Earl of Derby paid 'two shillings for two pounds of penydes' (similar to sticks of barley sugar). That's the equivalent of about £80 today.

Following the exploration and colonisation of the New World by Europeans, sugar cane plantations sprang up in the Caribbean and South America. Millions of enslaved workers, brought from Africa, harvested the crops, making the plantation owners and merchants wealthy and enabling sugar to become an affordable and plentiful commodity in Europe and elsewhere. Sugar consumption in Britain increased almost 22-fold in 200 years: from just under 2 kilograms per person per year in 1704 to 8 kilograms in 1800 and 19 kilograms (41 pounds) in 1901.

One of the problems with ordinary sugar is that it's high in calories so that we're likely to gain weight if we overindulge. Fortunately, some substances taste sweet even at very low concentrations so they can be used as low-caloric sugar substitutes. The first of these was discovered by accident in 1879. In a laboratory at Johns Hopkins University, the Russian chemist Constantin Fahlberg was working with a derivative of coal tar, benzoic sulphimide, when he noticed a sweet taste on his hand. He called the new substance saccharin and a few years later began to manufacture it in Germany as a sugar substitute.[1]

Saccharin first ran into controversy in 1906 because of concerns about food additives that were raised following the publication of Upton Sinclair's novel *The Jungle*. Head chemist at the US Department of Agriculture, Harvey Wiley, suggested there should be a saccharin ban but President Theodore Roosevelt would have none of it. Roosevelt, who saw the new sweetener as a shortcut to losing weight himself, declared: 'Anyone who says saccharin is injurious to health is an idiot.' And that was effectively the end of Wiley's career.

During the First World War, use of saccharin soared because sugar was in short supply. In the 1960s, it began to be promoted for its weight-loss benefits under trade names such as Sweet'N Low. Another scare followed the discovery that high doses of saccharin can cause bladder cancer in rats. A 1977 Act of Congress required that a cancer warning be put on all the product's packaging. But in 2000, after researchers found that saccharin is metabolised differently in humans than in rats, the need for the warning label was withdrawn.

Saccharin is 300 to 500 times sweeter than ordinary sugar, so that a tiny pill or sachet of it is enough to sweeten tea or coffee. Its main drawback is that it leaves a slightly metallic or bitter aftertaste. Alternatives include aspartame, often used in low-calorie soft drinks, and sucralose, the world's most commonly used sugar substitute.

The sweetest substance in nature may well be thaumatin, also called talin, which is actually a protein. It comes from seeds of the katemfe plant (*Thaumatococcus daniellii*), a species found in West Africa. When the fleshy part of the fruit is eaten, the thaumatin molecule binds to taste buds, causing a sweet sensation that slowly builds and leaves a lingering aftertaste. With a sweetness 2,000 times as intense as table sugar, a little goes a long way.

Even sweeter is neotame, with about 8,000 times the sweetening power of sucrose. Pop a gram of this in your mouth and it would be equivalent in sweetness to that of eight kilogram-bags of granulated sugar (but without any of the calories!).

Topping the list of the sweetest of sweet substances known are various derivatives of guanidine, so-called because it was first produced in the laboratory from guanine, an amino acid originally isolated from seabird droppings (guano). One of the products obtained from guanidine is sucronic acid, which is about 200,000 times sweeter than sucrose. Not to be out-done is another derivative guanidine, known as luguname, developed at the University of Lyon in 1996, and which, weight for weight, is between 220,000 and 300,000 times as sweet as the sugar you spoon into your tea.

CHAPTER 27

Sticky Problems

NEANDERTHALS HAVE BEEN given a bad rap. They're often portrayed as being dim, brutish members of the *Homo* family tree who died out because they weren't as smart as us at adapting to changing conditions. Yet, they built hearths and used fire, crafted clothes and jewellery, made art and buried their dead. They were also the first species, so far as we know, to invent a kind of glue – 200,000 years ago.

Many different natural materials will work as simple glues. Beeswax, resins from trees, and bitumen are among the sticky substances in nature that will help hold things together. Neanderthals knew about these materials and used them when they were available. But they also did something more. They found that if birch bark is heated under specific conditions, it produces a kind of tar, which can then be used to join stone tools to wooden handles.[1]

For the past few years, researchers from Leiden University and the Delft University of Technology, in the Netherlands, have been testing glues known to our Stone Age ancestors. They've found, from measurements on the effect heat has on the flow and hardness of the materials, that birch tar is superior to other natural adhesives. Their investigations have revealed not only how Neanderthals made birch tar glue but why they went to such trouble to fabricate it when other, simpler alternatives were available.

Some living things have been sticking to surfaces, as if they had glue on their feet, for many millions of years – long before humans or our close relatives came on the scene. Houseflies, along with many other insects, have no trouble in walking up windows or across ceilings without falling off. The same is true of some reptiles and amphibians, such as geckos and tree frogs. But none of these animals uses adhesives or even suction pads to hold on. A scanning electron microscope reveals lots of tiny hairs or bristles on the ends of the animals' feet. It used to be thought that these hairs gripped on to equally tiny footholds in the form of bumps and fissures that are present even on seemingly smooth surfaces like glass. But that's not how they work at all.

Zoom in on a gecko's foot and you'll see that its bulbous toes are covered in hundreds of miniature bristles called setae, each of which branches into hundreds of even smaller bristles called spatulae.[2] The bristles can get so close to the surface of a wall or overhang that molecules in the tiny hairs and on the surface interact by what's known as van der Waals force. This type of electrostatic bond, although weak, is more than strong enough, when multiplied across many thousands of spatulae, to counteract the downward force of gravity on the animal.

The setae are tremendously flexible and responsive. By rapidly varying the angle of the hairs to the surface and how much they extend, a gecko can stick and unstick its feet so fast that it can move up a vertical windowpane at up to twenty body lengths per second. Scientists have developed a mathematical model which shows how, if the setae are bent closer to the plane of the surface, the area the gecko can bind to increases and the animal can support more of its weight.

Geckos' feet can't stick to Teflon – but then what can? Also, if it's been raining, you may notice that geckos slip and slide as their feet fail to grip in the moisture.

Flies also have numerous setae on their feet. Each foot ends in two pads, called pulvilli, from which the setae spring. But the hairs on the feet of wall-climbing insects work in a different way – by producing a glue-like substance made of oils and sugars. A research team from the Max Planck Institute in Germany studied hundreds of species of insects and found that all of them leave a trail of sticky footprints.

A potential difficulty, if you use glue on your feet to hang from walls and ceilings, is how to unglue yourself every time you want to take a step. Look at a fly's foot through an electron microscope and you'll see that it comes equipped with a pair of claws. These can be used in various ways – peeling, pushing or twisting – to lift the sticky foot off a surface. By keeping at least four feet down at all times, a fly can run quickly, even upside-down, without falling off.

By 6000 BCE, Neolithic cave-dwellers near the Dead Sea were using collagen glue, obtained from the skin, sinews and cartilage of animals. They applied it not only as an adhesive for holding together tools and utensils but also as a lining in rope baskets and embroidered fabric, and even as a means of decorating skulls with crisscrossed patterns. The ancient Egyptians used and developed a variety of glues, both plant- and animal-based, from about 2000 BCE on. What's surprising though is that the Dead Sea cavemen, when it came to their knowledge of collagen glue, were more advanced. The Egyptians applied collagen in gelatinous form as furniture adhesive, as evident in chairs from the tombs of Pharaohs. Their Neolithic predecessors, however, supplemented their hide glue with plant-tissue additives in order to vary its texture according to the task at hand.

Throughout most of human history, sticking things together has involved various substances obtained from the animal and plant world. The first patent for an adhesive, derived from fish, was handed out in Britain in 1750. The next hundred years saw a plethora of different glues being made, patented and sold based on everything from rubber to milk. But it wasn't until the early part of the twentieth century that chemists began to develop synthetic adhesives. The first of these to be marketed commercially, in 1910, was made by the Swedish company Karlssons Klister.

As often happens in science, the discovery of what's become the most famous of all adhesives, Super Glue, came about by accident. It all started during World War II when Harry Coover, a young American chemist who'd just earned his PhD from Cornell University, was working for B. F. Goodrich Company as part of a team doing research into chemicals known as cyanoacrylates.[3] The team's goal was to make an optically clear plastic that could be used for precision military gunsights. The cyanoacrylates were clear enough but were also incredibly sticky, making them almost impossible to work with. As soon as they came into contact with moisture they bonded, hard and fast, to anything they touched. Obviously they were useless as a transparent plastic, but...

In 1951, Coover had moved to Eastman Kodak and was working at their chemical plant in Kingsport, Tennessee. Now, he and his group at Kodak had a different remit: to develop heat-resistant polymers for jet aircraft canopies. Coover remembered that the ultra-sticky chemicals he'd worked with at Goodrich would form tight bonds without any heat or pressure. Whatever items they tested in the lab became tightly and permanently bound together.

Coover and his employer realised they were on to something that went way beyond any specific application. A patent was granted to Coover for 'Alcohol-Catalyzed Cyanoacrylate Adhesive Compositions/Superglue'. Packaged by the company as 'Eastman 910', marketing of the product began in 1958 and the amazing adhesive was soon on sale everywhere under the sobriquet 'Super Glue'. Coover even appeared on the popular TV show *I've Got a Secret*, where he lifted the host, Garry Moore, off the ground using a single drop of his extraordinary chemical.

Most surprisingly, Super Glue found an important medical application. During the Vietnam War, field surgeons began using cyanoacrylate to treat war injuries. A quick spray of the substance over an open wound would immediately slow or stop the bleeding and buy time for wounded soldiers to be transported to where they could get conventional treatment. Many lives were saved in this way. Today, variants of Super Glue continue to be used in surgery. Often they're a preferred

A crane lifts a 17½-ton truck that is connected only by glue to aluminium cylinders projecting from the wheels.

alternative to sutures, for reconnecting arteries and veins, sealing bleeding ulcers, and closing open wounds with an improved cosmetic outcome.

A great range of powerful adhesives has now been developed. Judging their relative strengths is difficult because they work in different ways and under a great variety of different conditions. Still, if a winner has to be chosen, then it's probably a type of epoxy resin made by Delo, a German manufacturer of industrial adhesives. In 2019, its product broke the world record for the heaviest object ever supported simply by glue. Just 3 grams of the adhesive was applied to the end of an aluminium cylinder the size of a soft drink container. Having bonded with a 17.5-ton truck, the sticky cylinder, with truck attached, was lifted by a crane, leaving the vehicle suspended for an hour in mid-air.

CHAPTER 28

Phew

OUR SENSE OF smell – olfaction – is the oldest of the senses. Before there was sight, hearing or touch, there were primitive creatures, such as bacteria, that could respond to the chemicals around them. Smell is also unique in its ability to evoke deep-seated memories, even from early childhood. Yet, unlike sights and sounds, we often have difficulty describing specific smells in words. These differences are connected with the way olfactory information reaches the brain.

The pathways used by other senses, such as vision and hearing, start at the sense organs (in these cases, the eyes or ears) and then travel to a kind of neural relay station – the thalamus – before passing on to the rest of the brain. Smell information, in contrast, goes directly to the olfactory bulb, where it's processed, without going through the thalamic relay station. This direct connection between the outside world and the specific site of processing in the brain, neurologists believe, is why memories of smells can be so evocative yet hard to describe.

The molecules to which our sense of smell responds are airborne. They enter through the nose and mouth and attach to receptor cells that line mucous membranes at the back of the nose. Each of us has millions of such cells but, according to data gathered during the Human Genome Project, only about 400 different types.

When detectable molecules, or odorants, attach to receptor cells, small electrical impulses are generated. These impulses travel to the brain, which quickly, in about a tenth of a second, identifies the particular smell. With just a few hundred types of receptor cells it might seem we'd be limited in the number of different smells we could recognise. Each odour receptor, however, can detect not just a single, specific smell but a range of similar odorants. What's more, most odours evoke a response from more than one type of receptor. Because the number of combinations and permutations of olfactory receptors is huge, so is the number of different smells that we can detect and distinguish – perhaps as large as a trillion.

Smells range from delightful to disgusting, and people differ in their preferences. Partly, this is because which types of olfactory receptor cells are active in your nose is determined by your genes. Having said that, a study carried out by researchers at the University of Oxford and the Karolinska Institute in Sweden found that vanilla is the world's most universally appreciated scent.[1] More than 200 people from a variety of backgrounds, including many from non-Western indigenous groups, were asked to rank a range of smells from best to worst. Topping the list was vanilla, derived from the beans of a member of the orchid family, followed by chemicals that give peaches and lavender their pleasant odours.

At the other end of the olfactory scale are smells so bad it's almost impossible to tolerate them. Chemists have found that, in general, the stinkiest substances are ones in which the molecules are bigger and heavier, whereas pleasant smells tend to be associated with molecules that are more compact and lighter. An evolutionary factor is also at work: smells that are obnoxious to us, such as those of rotting food, often come from stuff that's bad for our health or survival. Laboratory rodents that, after generations of breeding have never come across a cat, still react with fear to cat odour but not to other new and noxious smells.

A handful of plants rank high on most people's list of smells to be avoided, if they've ever had the misfortune to catch a whiff of them. Titan arum (*Amorphophallus titanum*) boasts the largest unbranched inflorescence in the world. The flower is so huge – up to 3 metres tall – and takes so much energy for the plant to produce, that it appears only once every five to ten years. When it does, everyone around knows about it even with their eyes closed because Titan arum truly lives up to its alternative name, corpse flower.[2]

A specimen of Titan arum, or corpse flower, in bloom
at the New York Botanical Garden in 2018.

Equally appalling is the aroma of another so-called carrion plant, *Rafflesia arnoldii*, the aptly named stinking corpse lily, which attracts flies (to do the job of pollination) as much as it repels humans. It also blooms far more often than Titan arum, producing flowers that emit their whiff of decaying body for several days at a time.

Among fruits, the most notorious for smell is the durian, of which about nine species are edible. Those who've acquired a taste for it claim it has a pleasant fragrance and flavour. Its flesh is eaten at various stages of ripeness and the flavour was described by the nineteenth-century naturalist Alfred Wallace as like 'a rich custard highly flavoured with almonds'. But, for many, the aroma is so strong and evocative of everything bad, from rotten onions to raw sewage, that it overwhelms any desire to do a taste test. Some hotels and public transport services in Southeast Asia, where the fruit is used in a broad range of dishes, have banned durian for fear that it will turn away customers. And if the odour of fresh durian is bad enough (for non-aficionados) that of rotten durian is enough to cause a stampede. In 2018, 500 students at the University of Melbourne were evacuated after the alarm was raised of a possible gas leak. The culprit turned out to be a durian that had been left in a cupboard, gone well past its sell-by date and whose offensive bouquet had been widely disseminated through a building via the air conditioning system.

Of more conventional foods, cheese has a special reputation for its pungency. Limburger, from Germany and the Low Countries, has an aroma reminiscent of stinky feet thanks to the bacterium *Brevibacterium linens*, which is used to ferment a number of rind-washed soft cheeses and is also ubiquitous on human skin where it causes foot odour. But the stinkiest cheese of all, according to scientists at

Cranfield University, is Vieux-Boulogne, a soft cheese from northern France.[3] The Cranfield researchers employed an 'electronic nose', normally used to sniff for urinary infections and tuberculosis, but which registered positive when aimed at the pungent delicacy.

In the animal world, skunks are infamous for the powerful scent which they can spray up to 3 metres from their anal glands to ward off predators. Although most people react strongly to it, about 1 in 1,000 of the population have what's known as specific anosmia to skunk spray and can't sense it at all. Even smellier, by a factor of five to seven, is the defensive repellent fired out by the lesser anteater, or tamandua.

Among the chemicals in skunk and anteater spray is a thiol, an organic sulphur-containing compound. Thiols, also known as mercaptans, also help give onions and garlic their distinctive odours, and contribute to the smells of rotting meat and some cheeses.

There are thousands of different kinds of thiol and our noses are exquisitely sensitive to many of them. In the slightest intake of air, containing many billions of molecules, we can detect if just a few of them are thiols. If you 'smell gas', perhaps from a small leak in your home, what you're really smelling is not natural gas, which is odourless, but a tiny amount of thiol that has been added to the gas as an odorant to make it detectable.

Of all the substances on Earth, one is repeatedly cited as being among the very worst-smelling in existence. Thioacetone is another organosulphur compound, like the thiols but belonging to a different group known as thioketones.[4] It's an orangey-brown substance that above −20 °C converts to trithioacetone, the molecules of which contain six-membered rings of alternating carbon and sulphur atoms.

Two German chemists, Eugen Baumann and Emil Fromm, were the first to obtain thioacetone. Their attempts to distil it in 1889 in the city of Freiberg were followed by cases of nausea, vomiting and collapse within a 0.75-kilometre radius of the laboratory due purely to the smell, even though only a small quantity was involved. An 1890 report by chemists at the Whitehall Soap Works in Leeds described its stench as 'fearful' and that it seemed to get worse with dilution.

In 1967, researchers at the Esso Research Station in Abingdon, Oxfordshire, described their experience following the accidental release of a small amount of thioacetone. A stopper from a bottle containing the substance came loose and, although it was quickly replaced, the damage was done. The minute quantity of vapour released into the atmosphere quickly led to cases of nausea and sickness in a building 180 metres away. Just as the Whitehall Soap scientists had found, dilution seemed to make matters worse. Workers in the lab were mystified since they weren't troubled by the odour of the escaping chemical and at first denied they could have been responsible for the outbreak of illnesses much further away. An Esso article on the incident reported on further fallout: 'Two of our chemists who had done no more than investigate the cracking of minute amounts of trithioacetone found themselves the object of hostile stares in a restaurant and suffered the humiliation of having a waitress spray the area around them with a deodorant.'

CHAPTER 29

Next to Nothing

MANY YEARS AGO, my wife and I helped organise a balloon competition at our daughter's primary school to raise money for charity. For £1 you could buy a toy balloon filled with helium, attach your name and address to it and let it go. There was a prize for the person whose name tag was returned from the greatest distance. The results were fascinating. About two dozen tags were found and sent back with the location of where they'd landed. On a map, it turned out that all the balloons had been carried more or less south by the wind, from the release point near Carlisle. Some had travelled as far as London, about 420 kilometres away. Incredibly, three had crossed the English Channel into France and the winner made it as far as the Pyrenees on the border of France and Spain – a distance of more than 2,000 kilometres.

Balloons filled with helium rise because the density of helium (0.164 kilograms per cubic metre) is less than that of air (1.28 kg/m^3). Hydrogen and helium are the two lightest elements. Hydrogen has a lower density even than helium (0.082 kg/m^3) and was used in early airships, such as the great German Zeppelins. But it's also explosively inflammable and, following the *Hindenburg* disaster, was replaced in airships by helium, which is inert.

Hot-air balloons also rise, because the density of air that's been heated by a burner is slightly less than that of the surrounding atmosphere. The balloon's envelope has to be big enough so that the aggregate density of the envelope, the gas it contains and anything suspended from the balloon, such as a gondola with people in, is lower than the density of the air displaced.

Hydrogen, helium and hot air are all lighter than cold air. Other things that can appear to float in air are really just slowly falling or being kept up by the wind or other currents of air. A toy balloon filled with ordinary air falls because the weight of its rubber skin makes its overall density greater than that of the air it displaces.

Feathers drift gently down, but can be carried back up temporarily by a gust of wind, because they're light and have a large surface area like the canopy of a parachute. Wind-dispersed seeds and spores are able to stay aloft for extended periods for the same reason: their density is low – not much more, overall, than the surrounding air – and they often have structures that help them catch the breeze.

An incredible variety of microscopic living things can remain airborne long enough to travel great distances. Again, they're floating not in the sense of being buoyant but because they're small and light enough that air currents and thermal motions can, for a while at least, counteract the downward force of gravity. Wave a petri dish around, containing a growth medium, almost anywhere on the planet, then put it in an incubator. The result: you'll culture hundreds of species of soil, gut and faecal bacteria, fungi, and common viruses. We live constantly amidst an invisible sea of such microbes, which is why our immune system's first line of defence is a similarly diverse army of friendly microorganisms, both outside and inside our bodies.

Some larger creatures, too, are light enough to remain suspended for long periods of time and travel to great heights in the process. Outstanding among these are young spiders, which release gossamer threads to catch the wind and then rise and drift at the mercy of air currents.[1] The behaviour is called ballooning or kiting and, in extreme cases, can carry a spiderling hundreds of kilometres from its launch site and to heights of 5,000 metres.

Only animals that flap their wings have been recorded flying higher than ballooning insects. Butterflies have occasionally been observed at altitudes up to 6,000 metres. As for birds, bar-headed geese have been seen passing over the Himalayas at around 8,500 metres, while a Rüppell's griffon vulture was spotted at an altitude of more than 11,000 metres – the same as that of a cruising airliner.

But let's get back to Earth and talk about the lightest solids. After all, a bird, like a plane, is much heavier than the air it displaces so it normally has to put a lot of effort into staying aloft. Among the elements, the least dense solid one is lithium – a metal that comes third in line in the periodic table after hydrogen and helium. It's very reactive, though, so it's always combined with other substances in nature. One of the lightest rocks is pumice, familiar as a skin abrasive in bathrooms around the world. It's made during explosive volcanic eruptions when lava forms a froth containing masses of air bubbles, which are locked into the lava as it solidifies.

We – our bodies, that is – also contain a lot of air, especially when we breathe in. There's no danger of us floating away: the density of air is 1.2 kg/m^3, whereas we clock in at about $1,000 \text{ kg/m}^3$, more or less the same as the density of water. People who are obese are more buoyant in water than lean individuals because fat is less dense than muscle. But taking in a big lungful of air reduces the aggregate density so that, even if you're skinny, you'll displace at least your own weight of water and therefore be able to float.

On the slightly gruesome side, a similar explanation accounts for why dead bodies are often found on the surface of water. At first a body usually sinks. But, as it putrefies, microbes within the corpse release gases which decrease the overall density of the body and cause it to rise.

Any solid that contains a lot of air pockets will have a low density. The least dense of all are some artificial materials that have been developed not simply to be lightweight but because they have other extraordinary physical properties. One of these is aerogel, sometimes referred to as 'frozen smoke' because of its wispy appearance.[2] Aerogel is made from a porous, sponge-like structure of silicon dioxide – the same chemical as found in glass but in a form that's 1,000 times less dense. It was created in 1931 by Stanford University researcher Samuel Kistler as a result, according to one story, of a bet with fellow scientist Charles Learned over who could be first to replace the liquid in a jelly without causing it to shrink.

Weight for weight, silica aerogel is remarkably strong. A block the size of an average person weighs less than a pound yet could support a small car. It's also an extremely good thermal insulator, allowing less heat to pass through it than through an equal volume of air.

Today, different types of aerogel are known. The most astonishing of all is aerographene, in which the structural component consists of graphene – carbon arranged as a single layer of atoms, tightly bound in a hexagonal honey-comb lattice. Aerographene is currently the least dense solid substance on Earth. A future application could be mopping up oil spills. It can absorb up to 900 times its own weight in oil – and do it fast. One gram of aerographene can absorb

69 grams of oil every second. It's also super-strong: a block the size of your thumb could be supported by a single blade of grass yet is ten times stronger than steel.

Of course, the lightest substance in the universe isn't a substance at all but the absence of one – a vacuum. A balloon filled with a vacuum would rise much faster than one containing hydrogen or helium. The problem is, an ordinary balloon would collapse if you simply sucked out all the gas from it. A vacuum balloon would have to be rigid and, weight for weight, incredibly strong, so that the envelope could hold up against the outside air pressure. Substances such as aerographene bring us one step closer to such a tantalising possibility.

CHAPTER 30

Bubbles: Big, Beautiful and Bizarre

EVERY YEAR SINCE 1825 (except for the war years 1939–42), the Royal Institution in London has held a series of Christmas Lectures for children on a topic of scientific interest. In 1890, the lectures were given by physicist Charles Boys on the subject of soap bubbles. 'I hope', he said in his introduction, 'that none of you are yet tired with bubbles, because, as I hope we shall see during this week, there is more in a common bubble than those who have played with them generally imagine.'

The bubbles of our childhood are nothing more than a soap film wrapped around a pocket of air. Two layers of soap molecules form the inside and outside surfaces of the film and are separated by a thin layer of water. A bubble – until it bursts – is airtight. Even if it isn't popped on purpose or by touching something that breaks the film, it will burst spontaneously when the water between the layers of soap molecules evaporates. Blown on a cold winter's day, bubbles tend to last longer because the evaporation is slower and the bubbles may even freeze.

The key to understanding the shape of a bubble is surface tension – the force that acts on its surface like an elastic skin. Surface tension comes about because of the attractive force between molecules of a liquid. Inside a body of liquid a molecule is pulled equally on all sides by its neighbours and so experiences no overall force. But at the surface, molecules are pulled only sideways and downwards, which has the effect of making the surface appear as if it has a skin.

It's a common misunderstanding to suppose that the reason you can't blow bubbles with water alone is that the surface tension of water is too low and that soap increases it. In fact, the opposite is true. Adding soap *decreases* the surface tension. Water bubbles burst almost the instant they form, for a couple of reasons: the surface tension is too great, causing them to tear themselves apart, and evaporation from the surface of the bubble makes them too thin so that they pop. Soap molecules help bubble formation because each consists of a chain of atoms (carbon and hydrogen) one end of which is hydrophilic ('water-loving') and the other hydrophobic ('water-hating'). In a soap-and-water solution, the hydrophobic ends of soap molecules move as far away as possible from the water and so end up at either the inner or the outer surface of a bubble. Meanwhile, the hydrophilic ends project into the water sandwiched between the two layers of soap molecules, which increases the separation of the water molecules and thereby reduces the attractive force between them. Result: the surface tension is lowered. What's more, because the water is partly protected by the soap films it evaporates more slowly.

If kept airborne, bubbles typically last ten to twenty seconds before they burst. But their lifetime can be greatly extended by keeping them in a sealed container in which the air is saturated with water vapour to reduce the rate of evaporation. Eiffel Plasterer, from Huntingdon, Indiana, became fascinated with bubbles as a physics teacher in the 1920s and went on to become well known for his bubble-making shows and demonstrations. His prime-time TV appearances included one on *Late Night with David Letterman* in which he managed to form a complete soap bubble around the host. He also holds the world record for soap bubble longevity, having once blown a bubble in a sealed jar that lasted for more than eleven months.[1]

Bubbleology attracts more than its fair share of enthusiasts and master exponents who constantly vie to outdo each other. Matěj Kodeš of the Czech Republic currently tops the list for enclosing the largest number of people within a single bubble – 275. He also managed, in 2010, to form a bubble around a 6-metre-long truck. Canadian citizen Fan Yang is distinguished for having blown the largest number of bubbles – twelve – one inside another, like a Russian doll. Samsam the Bubble Man (aka Sam Heath) of the UK boasts three records: the most bounces by a bubble (38), the longest chain made from interlocking bubbles (26), and the largest frozen soap bubble, with a volume of 4,315 cubic centimetres. As for making the largest-ever free-floating soap bubble, that honour goes to American Gary Pearlman who, in 2015, created a monster with a volume of about 96.2 cubic metres.

The largest soap bubbles are more than 10 metres wide.

Giant bubbles wobble like an underset jelly on the tray of a nervous waiter. Small ones, however, maintain a constant shape, which, as everyone knows, is a sphere. No other shape encloses a given volume with a smaller surface area. Like all things in nature, bubbles tend towards the lowest energy state possible. In doing so they minimise the forces of tension in the soap film, which, in turn, minimises the surface area that encloses a given volume. The logic, and physics, behind why bubbles are spheres isn't hard to follow. But proving, mathematically, that the sphere is the surface with the minimum area for a given volume is surprisingly hard. In fact a complete proof came as recently as 1884.

In the nineteenth century, Belgian physicist Joseph Plateau came up with a number of laws that could be applied to the shape of bubbles. The first two are that soap films are made of smooth surfaces and that the average curvature is constant throughout each individual film. His third law states that when three soap films meet, they always do so at an angle of 120 degrees. This rule applies to when two bubbles are joined because, in this case, there's a meeting of three soap films: one for each bubble, and one for the boundary separating them. If one of the bubbles is larger, the boundary will curve inward towards it in order to satisfy the third law. Plateau's fourth law is that where three faces meet at 120 degrees, the edges of these faces will themselves meet in fours with angles of approximately 109.5 degrees – the tetrahedral angle. It's so called because if you draw lines from the corners of a tetrahedron to the centre, this is the angle at which they'll meet.

Plateau's results are all the more impressive because he figured them out when he was completely blind. What led to his loss of sight isn't certain but was probably connected with his tendency to perform risky optical experiments. He's known, for instance, to have once stared directly at the Sun for twenty-five seconds, to see what impressions it would leave on his retina.

So far we've only talked about soap bubbles. But bubbles are just globules of one substance inside another. You can get bubbles of air in water or any other liquid, including molten rock. When lava, thrown out of volcanoes, mixes with air and cools, it can form pumice, a type of stone that's riddled with tiny air cavities. Bubbles are sometimes seen in amber, which is fossilised resin from trees in which prehistoric insects are sometimes trapped. A piece of amber found in the Dominican Republic, and thought to be 20 to 30 million years old, contains a pocket of water in which there's an even smaller bubble of air.[2] If this extraordinary specimen is moved, it causes the little air bubble – captured long before anything human walked the Earth – to slide back and forth.

At the other end of the size scale are immense bubbles in space. One of these is known as NGC 7635 or, more picturesquely, the Bubble Nebula. It lies about 10,000 light-years away and has been formed by an intense stellar wind – a fierce outflow of energetic particles – from young, hot stars, forty-five times more massive than the Sun. The stellar wind, moving at over 6 million kilometres per hour, sweeps up cold, interstellar gas in front of it, forming the outer edge of the bubble.

Much bigger, but not as round, is the so-called Local Bubble – a region of space, at least 1,000 light-years across, that contains the Solar System and many of our near stellar neighbours and star clusters. The density of gas in the Local Bubble is much lower than that of the interstellar medium outside. This immense cavity, in which the Sun lies, is thought to be the result of a series of supernovae that exploded within the past 10 to 20 million years.

CHAPTER 31

Acid Test

EVERY SECOND OF every day we're in contact with a strong acid of which – most of the time – we're blissfully unaware. Gastric acid is released into the stomach from glands in the stomach wall and plays a vital role in digesting our food and killing harmful bacteria before they can enter the intestines. It produces a stinging sensation if it gets into your throat, when you retch or vomit, and would cause redness, blistering and, eventually, a nasty burn if left on exposed skin for several hours.

Chemists measure the strength of acids and alkalis by their pH or 'potential hydrogen' value. Pure water is neutral with a pH of 7, while orange juice is mildly acidic with a pH between about 4 and 3.5. Vinegar, which is mostly acetic acid in water, scores a pH of about 2.5. A powerful and dangerous acid, such as that in a lead-acid car battery, has a pH of less than 1. To accomplish its task of breaking down proteins and fibrous plant material, gastric acid has a pH level that varies between about 2 and 1.5.

This begs the question, if gastric acid is strong enough to decompose the proteins in meat and can burn skin why doesn't the stomach digest itself?[1] The answer lies in the protective coating of mucous and bicarbonate, or 'antacid', which cells in the stomach's lining produce. Having said that, if the pH level of the gastric juice stays too high for too long, a person will feel pain and could, over a period of time, develop a stomach ulcer.

Acidity was originally recognised as a distinct property of some substances because of their taste. In Latin, the word for 'vinegar' or 'the tang of vinegar' is *acetum*, which became anglicised as 'acid'. Lemon juice, strong black tea and tart or sour wine are among common liquids described as being acidic on the basis of how they taste. But all of them contain just weak *organic* (carbon-based) acids – citric, tannic, tartaric and so on – which aren't harmful to us. The last thing you'd want to do with a strong acid is put it on your tongue.

Alchemists in the Middle Ages learned how to make *inorganic* or 'mineral' acids. Spearheading this effort was a mysterious character known as Geber, who lived in the fourteenth century and was probably Spanish. Geber wasn't his real name, but one taken from the much earlier Arabic alchemist Jabir ibn Hayyan (Geber being a Latinised form of Jabir). In several influential books, Geber encapsulated the chemical knowledge of that period, including how to make a variety of strong acids capable of dissolving metals. This information was of enormous interest to alchemists because one of their ultimate goals was to be able to turn 'base' metals, such as iron, into gold.

Geber gave the first clear instructions on how to make 'oil of vitriol' – what we now know as sulphuric acid. He described heating 'vitriols' (certain sulphur-containing compounds) to produce sulphur and a gas, which, when cooled, would condense with water vapour, to form a liquid that could dissolve metals. He described a similar process, using nitre (potassium nitrate) and vitriols, to make *aqua fortis* (nitric acid) that would break down silver. Most intriguing of all to alchemists was his explanation of how mixing sal ammoniac (ammonium chloride) with *aqua fortis* produced *aqua regia* (a combination of nitric and hydrochloric acids). *Aqua regia* – 'royal water' – held the distinction of being able to dissolve even gold.

Alchemists believed that to transform a base metal into gold they first had to 'unmake' that metal. So, one of the things they sought was a universal solvent – a liquid capable of dissolving any other substance. It seemed to them that *aqua regia*, with its power to dissolve gold, might be such a solvent, or at least a good approximation of one. If it could 'unmake' gold, the notion went, perhaps the process could be reversed and gold could be created from other substances.

Geber supported the older idea, held by Arabic alchemists, that all metals, including gold, were composed of sulphur and mercury. Only the proportions differed from one metal to another, he taught, so that by changing these proportions it seemed that one metal could be transformed into another. The alchemists didn't know that metals such as gold, silver, iron, tin and lead were elements in themselves incapable of being broken down into simpler substances.

Sadly, belief in the transmutation of base metals into gold lingered long enough for swindlers to take advantage of it. One trick involved placing a silver-coated gold coin in nitric acid. The silver would be dissolved leaving the gold underneath, which the acid couldn't attack, exposed. Unwitting wealthy patrons would be persuaded into thinking the silver had been turned into gold and thereby tricked into handing over money in order to make more gold. Of course, after that they'd never see the swindlers, or their money, again.

Starting in the second half of the eighteenth century, true chemists, such as Antoine Lavoisier in France and Humphry Davy in England, brought a new level of understanding about the nature of acids and their chemical opposites, bases. In time it became clear that acids were distinguished by their ability to add positive hydrogen ions (protons) to water. Hydrochloric acid (HCl), for instance, releases hydrogen ions (H^+) and chloride ions (Cl^-). A base, by contrast, releases hydroxyl ions (OH^-), which combine with hydrogen ions to give water molecules. This is why bases can neutralise acids.

One way to measure the strength of an acid is by how readily it produces hydrogen ions. Strong acids are ones that completely dissociate – break apart into ions – in solution. But just how strong can acids get?

So-called superacids are even stronger than traditional mineral acids – stronger than 100 per cent pure sulphuric acid. Think of the explosive reactions or nasty burns that ordinary strong acids are capable of causing, then multiply those effects many times. The pH scale is useless for measuring superacids because they react too violently with water, and the way they attack other substances isn't simply by releasing hydrogen ions.

Instead, chemists use a special way of gauging the strength of superacids called the Hammett acidity function, H_0. On this scale, pure sulphuric acid scores −12. In 1927, it was found that perchloric acid, with the formula $HClO_4$, was stronger than sulphuric, having an H_0 of −13 (the more negative, the stronger the acid). But two other substances, also known at that time, proved to be even stronger. Fluorosulphuric acid (HSO_3F) turned out to be about 1,000 times stronger than the most concentrated sulphuric acid, and antimony pentafluoride (SbF_5), first reported in 1904, was especially dangerous in the violence of its reactions with other substances, including human skin.[2]

Some of the strongest of all superacids have been made by combining other acids. In the 1960s, a group of chemists at Case Western Reserve University, in Cleveland, Ohio, led by George Olah, teamed fluorosulphuric acid with antimony fluoride. After a Christmas party in 1966, a member of Olah's lab put a wax candle into the new acid concoction and saw it rapidly dissolve. The superacid combo had cleaved the paraffin chain of carbon atoms in the wax in a reaction never before witnessed. Olah's team were so struck by the discovery that they dubbed their wax-dissolving brew 'magic acid' – a name that also appeared in the scientific paper they published on their discovery. For the record, the Hammett acidity of magic acid is −23, dwarfing that of any acid you might find on a typical laboratory bench. It's a hundred billion times stronger than the purest sulphuric acid.

But we're not quite done. The most powerful acid known at present puts even magic acid in the shade. Mix hydrogen fluoride and antimony pentafluoride and you'll create fluoro-antimonic acid.[3] Whether you'd survive the process, however, unless you know exactly how to work safely with such deadly substances, is debatable. One of the problems is finding something in which to contain the product. Fluoroantimonic acid is so reactive it attacks all metals, most organic substances, including skin and bones, plastics and even glass. Teflon is one of the few materials that will hold this supremo of superacids at bay. It registers up to −28 on the Hammett scale, making it as much as 100,000 times stronger than magic acid and 10 quadrillion times the strength of sulphuric acid.

CHAPTER 32

Clearly the Best

IN THE 1986 film *Star Trek IV: The Voyage Home*, chief engineer Montgomery Scott gives away the formula for a yet-to-be-invented substance – 'transparent aluminum' – to a twentieth-century scientist. This enables see-through, light-weight but super-strong walls to be built for a tank to carry whales on board a spacecraft. Fiction has now become reality – almost. Aluminium oxynitride, or ALON, is a ceramic compound, containing aluminium, oxygen and nitrogen, that's more than 80 per cent transparent to visible light. It's also tough enough to withstand blasts and bullets and can be fabricated into windows, domes, tubes and other forms.

We're surrounded by substances that are transparent – most obviously, air and water. Other natural materials we can see through clearly include rock crystal quartz, fine quality diamonds and corundum. The first transparent material to be made artificially was glass, using silica (silicon dioxide), found in sand, as the main ingredient.

Glass-making goes back thousands of years to its origins in the Middle East. Early on, its products were purely orna-mental – jewellery and *objets d'art*. The Romans were the first to make glass windowpanes but, although these let through light, they were coarse, unevenly thick and impossible to see through clearly. Only as more sophisticated manufacturing techniques evolved, in mediaeval Europe, were glass windows made that were truly transparent.

One of the strangest consequences of improved glass-making was the delusion that one's own body was made of glass and could shatter into pieces unless treated with great care. King Charles VII of France was famously afflicted with this disorder and wore clothing reinforced with iron rods to protect his supposedly fragile body. As a further precaution, he forbade anyone, including his closest advisors, to come near him. In another case, a man became convinced he had glass buttocks which could smash into shards if he sat down. He also feared that if he left his house, a glazier might come along and melt him down into a windowpane.

Today, it's hard to imagine a world without glass. Its key property of letting light through without absorbing or scattering it is invaluable, not only in the windows of buildings and all forms of transport but in eyeglasses and other optical devices.

Imagine a stream of light particles – photons – arriving at the surface of a substance. The electrons that are bound to the atoms making up the substance occupy certain specific energy levels. A photon will be absorbed if it meets an electron and is able to bump it to a higher level. The difference between the original level and the new, 'excited' state is known as an energy gap. If the energy gaps are bigger than the amount of energy possessed by the incoming photons, the light can't be absorbed and will continue on its journey.

But to be transparent a substance must not only avoid absorbing photons, it mustn't scatter them either. If the photons are scattered in different directions along the way, the substance will be merely translucent. Scattering happens if the light suddenly encounters regions within a material of different density or if it comes to a so-called grain boundary where two different microscopic crystals meet.[1]

Modern clear glass has a very uniform composition and therefore a high transparency. But the thicker it is, the less light it allows through. A 3-millimetre sheet of ordinary window glass lets about 91 per cent of the light falling on it to pass through to the other side. Double the thickness, to 6 millimetres, and the amount transmitted is 91 per cent of 91 per cent, or 83 per cent, and so on. If a sheet could be made that was 1 metre thick and free from impurities or imperfections, the amount of light making it all the way to the other side would be just 0.002 per cent. If full daylight fell on one side, it would be as dim as a moonlit night on the other side.

We know of several substances that have a higher transparency than glass. Pure quartz crystal is one of them. Among artificial materials, the best known are the transparent acrylics, used for everything from geodesic domes to the shielding around a hockey rink that prevents the puck from flying into the crowd. ETFE (ethylene tetrafluoroethylene) and PMMA (polymethyl methacrylate), known commercially as Plexiglas, are both significantly more transparent than glass. They can also be made much thicker and still transmit most of the light that enters them.

Because of tiny imperfections in atomic arrangement, and physical limits, nothing is 100 per cent transparent except a perfect vacuum. Yet that hasn't stopped writers from speculating about the possibility of a substance that's completely invisible or scientists from trying to devise ways to create the illusion of invisibility.

In *The Invisible Man* by H. G. Wells, first published in 1897, the protagonist invents a chemical means of changing the body's refractive index so that it's the same as that of air.

He applies the chemicals to himself but then finds that he can't reverse the process. The Russian writer Yakov Perelman pointed out a major drawback of having a body that can't absorb any light at all: the invisible person would be blind because their retinas would let light clean through instead of absorbing it and then sending signals to the brain.

In the universe of *Star Trek*, some starships, such as those of the Romulans, carry a 'cloaking device' – a form of stealth technology that bends light and other forms of energy around the spacecraft so that it can't be seen. As in so many areas of human endeavour, the gap between science fiction and reality is rapidly narrowing. Research is progressing along several fronts to develop a form of cloaking device that actually works, for military and other purposes.

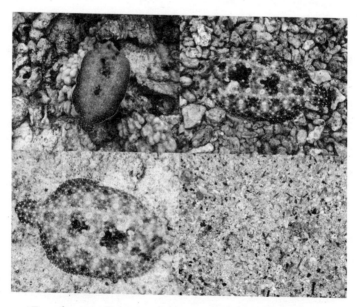

Four frames of a peacock flounder showing its ability to match its coloration to the seabed around and beneath it.

One approach to cloaking is via metamaterials that have properties not found in naturally occurring substances. It's possible, for instance, to design the optical parameters of a cloak so that it guides light around an object, rendering it invisible over a certain band of wavelengths.

A different strategy, called active camouflage or crypsis, makes an object seem to disappear by rapidly and accurately blending it in with its surroundings. To an observer it appears as if the object has vanished because it looks exactly the same as its background. A variety of marine cephalopods, such as squid, octopus and cuttlefish, as well as some reptiles and fish, use active camouflage by changing their colour or generating light from their bodies to match their background through bioluminescence.

Various systems are being developed to achieve active camouflage of objects and people. Defence organisations are especially interested because it provides a way to perfectly conceal equipment and personnel from visual detection. Early efforts in this direction were made during World War II based on a principle called diffused lighting camouflage discovered by the Canadian scientist Edmund Burr in 1940. Burr's idea was to project dim light onto the sides of a ship at night to match its background – the night sky – and thus hide it from German U-boats in the Battle of the Atlantic. The light would come from projectors mounted on supports attached to the hull, and the brightness would be controlled automatically using a photocell. A research project began in 1941 and tests were carried out on Royal Canadian Navy corvettes. The Royal Navy and US Navy also ran trials on the concept between 1941 and 1943 but it was never put into production.

Today, building on Burr's principles, a much more technologically sophisticated scheme is being developed. It uses cameras to sense the background and special coatings or panels to recreate the appearance of that background, even as it changes, moment by moment, on the side of the object or person to be camouflaged.

In one version of this approach, the goal is to make a military vehicle, such as a tank, invisible to infrared sensors. This is achieved by covering the side of the vehicle with what are called Peltier panels. Controlled by an electric current, these panels can rapidly heat up or cool down to match exactly the temperature of the background using information fed in real time from a series of infrared cameras. To an enemy's thermal imaging system, from a distance of more than a few hundred metres, the vehicle would be invisible.

The same technology can be used to achieve 'mimesis' – in other words, creating the illusion that the object being camouflaged is something other than what it really is. In this way, a tank might be made to look like a car or a rock. To achieve this illusion the Peltier panels would be fed with the infrared profile of the desired fake object from the cloaking system's library. Part of the panels would mimic the illusory object while part would be in cryptic mode, imitating the natural background.

There may be no such thing as a substance that's completely transparent. And a cloak of invisibility such as that worn by Harry Potter may forever be beyond our grasp. But clever use of technology can bring us as close as needs be to both these ideals.

CHAPTER 33

Rare

GOLD AND DIAMONDS are valuable because they're both beautiful and hard to find. But they're far from being the rarest substances on Earth. Gold isn't even the rarest metal. That distinction goes to rhenium, with an average concentration in Earth's crust of about one part per billion. Unsurprisingly, it was the last stable element to be discovered, in 1925, and is named after the river Rhine (Gaulish *rēnos*), from which the earliest samples were obtained. Total world production is only 40 or 50 tons per year, most of it being added to superalloys to make jet engine parts.

Almost as uncommon is rhodium, which was discovered serendipitously by the English chemist and physicist William Hyde Wollaston. On Christmas Eve, 1800, Wollaston and a wealthy colleague, Smithson Tennant, took shipment of 400 pounds of nearly pure platinum ore that had been smuggled out of the Spanish colony of Nueva Granada (now Colombia) via Kingston, Jamaica. The cost was a hefty £795 – the equivalent of about £85,000 today. But it was money well spent. In his backyard laboratory, Wollaston used the ore to develop a process to make malleable platinum metal. He kept the details of the process secret for nearly twenty years, until shortly before his death, and made huge profits by being the sole supplier of platinum in England.

While working with the ore, Wollaston and Tennant found four previously unknown metals – two each – with chemical properties similar to those of platinum. Wollaston discovered palladium (named after the asteroid Pallas, which had also been newly found) in 1802, and rhodium in 1803. *Rhodon* is Greek for 'rose' – like the distinctive red colour of the salts that turned out to contain rhodium. (Tennant separately discovered iridium and osmium in solutions obtained from the ore.)

Most of the rhodium produced every year is used in catalytic converters for cars or as a catalyst in the chemical industry. It's ultra-shininess and scarcity make it attractive for use in jewellery but only for those who can afford it. The price of gold, at the time of writing, is $1,819 per ounce; the same amount of rhodium will set you back $11,500.

Scarcity is a big factor in determining value, and pink diamonds are among the rarest types of gemstones in the world. One of them, known as the Pink Star, sold for a record-breaking $71 million at a Sotheby's auction in Hong Kong in 2017. Weight for weight, blue diamonds can sell for even more. Of course, the highest quality gems are also valuable because of their perfection and allure.

Some minerals are far less common than the likes of diamonds, emeralds and amethysts but are little known because they lack the same showy appearance. A good example is painite, discovered in the 1950s and at first thought to be a type of ruby, which itself is a form of corundum (aluminium oxide) with chromium impurities. But painite has a complex chemical makeup, which includes boron, calcium, zirconium, aluminium and oxygen. Its brownish-red colour comes from trace amounts of chromium and vanadium. Painite has been found in only one very small region of Myanmar, its rarity due to the fact that boron and zirconium hardly ever combine in nature. Up until 2004, only two crystals of painite had been cut into faceted gemstones, although recent discoveries in the same area have made several thousand new specimens available.

That leaves one other mineral unchallenged as the rarest known of all the 6,000 or so species identified on Earth. Just a single specimen of kyawthuite (named after Dr Kyaw Thu, formerly a geologist at Yangon University), a compound of bismuth, antimony and oxygen, has ever been found. It came from near Mogok, Myanmar – the same region that has yielded painite and also many rubies, sapphires and other varieties of semi-precious gems. This solitary specimen, now cut, facetted, and weighing a mere 1.6 carats (0.3 grams), is stored in the Natural History Museum of Los Angeles County.

Fewer and further between than anything mentioned so far are some unstable substances that break apart almost as soon as they form. The element astatine is in the same group of elements as chlorine, bromine and iodine – the halogens – but unlike them is radioactive in all its forms. In fact its name comes from the Greek *astastos* meaning 'unsettled'.

Astatine occurs naturally on Earth as a product of the decay of heavier radioactive elements. It has an atomic number of 85, which refers to the fact that there are 85 protons in its nucleus. Atoms of the same element all have the same number of protons but different *isotopes* of an element have different numbers of neutrons. Even the most stable isotope of astatine – astatine-210 – has a half-life of just over eight hours. If you were to gather all the atoms of astatine in Earth's crust together at any one time, you'd have less than half a gram of it – and half of that would have decayed into other elements within a few hours. Your efforts would be further hampered because even a tiny sample of astatine would be immediately vaporised by the heat of its own radioactivity.[1]

Astatine is the rarest natural element on the planet for the simple reason that it's so unstable. By the opposite token, the rarest *event* ever seen on Earth is the decay of an incredibly *stable* element. As we saw in Chapter 2, an unusual experiment set up underground in Italy to look for dark matter has detected the occasional breakup of individual nuclei of xenon-124, which has a half-life of 18 billion trillion years. Half-life is a measure of how long, on average, it takes for a particular type of nucleus to decay. Because radioactive decay is a purely random process, any given nucleus could decay a lot quicker or a lot slower than that. Xenon-124 holds the record for the longest half-life directly measured for any unstable isotope.

In the universe at large, some things we value highly on our own little ball of rock turn out to be not so rare after all. About 900 light-years away, in the constellation Aquarius, lies an 11-billion-year-old object the size of Earth. Astronomers recognise it as a white dwarf – the kind of burned-out remains of a star that the Sun will eventually become when it stops making light and heat by nuclear reactions. But this white dwarf is unusual. So cool and dim has it become, due to its great age, that it can no longer be seen. We're only aware of its presence because of the effect it has on the normally steady radio pulses coming from an equally unusual companion star – a pulsar.

The white dwarf, although invisible to astronomers, is revealed by the way its gravity periodically delays the signals from the pulsar. Over time, researchers have figured out the dwarf's distance, mass and age. It's slightly heavier than the Sun and composed almost entirely of carbon compressed into a planet-sized ball. It's also one of the coolest white dwarfs known, with a temperature of only about 2,800 °C – cool enough for its carbon to have crystallised. Put simply, this defunct star, a stellar corpse with the mass of the Sun and the size of Earth, is made entirely of diamond.

TECHNOLOGY

CHAPTER 34

Fastest Computer

I WAS LUCKY enough to have met, and worked at the same company as, the greatest designer of fast computers, Seymour Cray. The Cray-1, first announced in 1975, was the most successful and well known of the early supercomputers. But how does it stack up against high-end machines today or, for that matter, even a smart phone?

No area of technology has advanced more rapidly in an average human lifetime than computing. One of the earliest electronic computers, called Colossus, helped give Allied forces an important advantage towards the end of World War II. Built in 1944 at Britain's code-cracking headquarters at Bletchley Park, Colossus was designed specifically to break a top-secret cipher used to send messages between the German High Command and its army units throughout occupied Europe. By the end of the war there were ten Colossus computers at various locations in England, each weighing more than a ton and operated by a team of several dozen people.[1]

Colossus was fed information from a paper tape reader at a speed of about 5,000 characters per second. It used 2,400 vacuum tubes and was the world's first digital electronic computer. But it could do only one job: code-breaking.

The first *general-purpose* programmable electronic computer, ENIAC, was also built with a military purpose in mind. Completed in 1945 at the University of Pennsylvania, it was designed and used mainly to calculate artillery-firing tables for the United States Army's Ballistic Research Laboratory.

ENIAC was immense. When the press eventually learned of its existence, it was dubbed a 'giant brain' and became the archetype for futuristic computers of 1950s science fiction, which were envisaged as being colossal both in size and in their power to surpass human mental speed and intelligence. ENIAC occupied most of the 15-by-9-metre basement of Penn University's Moore School of Electrical Engineering. Its forty U-shaped panels, each about 0.6 metre wide, 0.6 metre deep and 2.4 metres high, were arranged along three walls. Boasting 17,000 vacuum tubes, 70,000 resistors and 10,000 capacitors, it could carry out up to 5,000 instructions per second (IPS) and, in the process, generated 174 kilowatts of heat.

Upon completion, ENIAC had cost the US government today's equivalent of about $6 million, but the war it was intended to help win was over. In fact, the first practical task to which it was put was performing calculations for America's hydrogen bomb project. In 1947, it was moved to the Army's Aberdeen Proving Ground in Maryland, where it ran continuously until 1955. A couple of other 'giant brains', also built for the US military, overtook ENIAC in terms of speed during the 1950s. First, in 1951, Whirlwind I, at the Massachusetts Institute of Technology, maxed out at 20,000 IPS. Then, six years later, the AN/FSQ-7, or Q7, a command and control system for air defence during the Cold War, lifted the high bar up to 75,000 IPS.

By this time, more and more uses were being conceived for computers, especially in science and business. The early 1960s saw the dawn of a new generation of computers based not on vacuum tubes, which were bulky, power-hungry and prone to burning out, but solid-state transistors. In 1960, IBM's transistorised 7090 series became available – at an equivalent price today of $20 million. Like many cutting-edge computers, then and now, it was initially installed at a research establishment – in this case, Los Alamos Scientific Laboratory. Its top speed, of 229 kIPS was just below that of Remington Rand's UNIVAC, the first of which was placed in Lawrence Livermore National Laboratory in the same year. In 1961, the IBM 7030, also known as Stretch, became the first computer to break the 1 million instructions per second barrier, although its 1.2 MIPS top speed fell well below the target that had been originally set.

What is often considered the world's first successful super-computer, the CDC 6600, made by Control Data Corporation under its chief designer, Seymour Cray, outpaced all competitors from 1964 to 1969. But it paled in comparison with its successor, the Cray-designed CDC 7600, which could hit 15 MIPS or, to use another measure of computer speed, 36 MFLOPS. 'MFLOP' stands for 'million floating-point operations per second' and is a commonly used unit in the field of scientific data processing.

In turn, the CDC 7600 was surpassed by Cray's third major supercomputer design, the Cray-1, first installed at Los Alamos in 1976. The Cray-1 was fast for a number of reasons: it used speedy chips; it exploited a novel kind of architecture known as vector processing; and it was compact, thereby minimising the distance that signals had to travel between different parts of the computer. Far from being the 'giant brain' of early SF tales, the Cray-1 was small enough to have fitted in the corner of a living room and, in fact, even looked like a piece of furniture, with seating, atop its power supplies, arranged around a C-shaped central cabinet.

For its time, the Cray-1 had impressive specs: a top speed of 160 MFLOPS, storage of up to 303 megabytes, and a memory (RAM) of 8 megabytes. The base price of a Cray-1 was about $8 million – equivalent to around $36 million today. That was a huge sum of money for a small company and I remember, in my early days with Cray Research, we held a celebration party after each sale.

By current standards, though, the Cray-1 appears comically underpowered. Compare it with a high-end mobile phone – say the iPhone 13 Pro Max, which costs about $1,300. The iPhone can perform 732 GFLOPS (732 billion floating point operations per second), and has 1 terabyte (1 trillion bytes) of storage and 6 gigabytes (6 billion bytes) of RAM. That makes it 4,800 times faster than a Cray-1, with 3,300 times more main storage, and 750 times more RAM. And at an asking price that's about 2,000 times less!

So is an iPhone a supercomputer? By 1970s and 1980s standards, yes. But, by definition, a supercomputer is among the fastest computers *of its day*, and today's supercomputers are very much more powerful than any contemporary mobile phone. The current champion is called Frontier and has been installed at Oak Ridge National Laboratory in Tennessee.[2] It's the world's first 'exascale' computer, which means it can perform more than 1 million trillion (10^{18}) FLOPS, or 1 exa-FLOP. Built at a cost of $600 million it replaced the previous record-holder, Japan's Fugaku, in 2022.

Like all of the world's fastest computers for the past several decades, Frontier consists of multiple high-speed processors and memory units working together in harmony. The result is known as a computer cluster – effectively a set of computers working together so closely that they form a single computer system with an architecture that's described as being 'massively parallel'. To enable such a system, with multiple interlinked hardware units, to work efficiently on a problem, has called for tremendous developments in software. Thanks to such developments, a task given to a machine like Frontier – for example, running a complex climate simulation program – can be broken down into many separate calculations which can all be carried out at the same time.

Already, computers more powerful than Frontier are being installed or developed. The first to overtake it will be Aurora at Argonne National Laboratory, on the outskirts of Chicago. By late 2022, Aurora's more than 9,000 computing nodes, spread across 200 cabinets, were ramping up towards a speed of about 2 exaFLOPS. Among its uses will be research on low carbon technologies, helping develop new materials for use in batteries and more efficient solar cells, and simulations in high-energy particle physics, cosmology and medicine.

The Exascale-class HPE Cray EX Supercomputer, known as Frontier, at Oak Ridge National Laboratory – currently the world's most powerful computer.

In the East, rival supercomputer pioneers China and Japan are in hot pursuit of the number one slot. Later this decade, and into the 2030s, computers capable of tens and even hundreds of exaFLOPS will emerge. The graph of computing power shows exponential growth so that by around 2036, the first zettaFLOP (1 billion trillion FLOP) machine is expected. Beyond that, in the early 2050s, the yottaFLOP milestone may be crossed. A yottaFLOP supercomputer would be a million times faster than one operating in the exaFLOP range. It could run a simulation that would take about six months using Frontier in just fifteen seconds.

In a separate development, researchers are starting to build quantum computers. Unlike classical computers, these operate using the strange rules of quantum mechanics. Whereas an ordinary computer works with 'bits' – binary digits – quantum computers deal in qubits, or quantum digits, which are linked together by a bizarre phenomenon known as quantum entanglement. An upshot of this is that, while carrying out a calculation that involves multiple possible outcomes, a quantum computer effectively splits into multiple copies of itself, so that all the outcomes can be explored simultaneously.

Quantum computers will eventually be able to far outperform their classical counterparts for certain applications, such as finding the prime factors of a very large number (a key problem in cryptography) or the optimal path between two points. One of the goals of quantum computer scientists is to demonstrate 'quantum supremacy' – situations where a quantum device can solve a problem that no classical computer can solve in a feasible amount of time. Several claims of quantum supremacy have already been made. For example in 2020, a group of scientists at the University of Science and Technology of China ran a physics problem on their quantum computer, Jiuzhang, which took twenty seconds to solve. A classical supercomputer, they estimated, would have required 600 million years.

If there's one thing the progress of computing has taught us, it's to expect the unexpected. Who, in 1970, could have predicted that within half a century or so we'd all be carrying around a device with vastly more processing power than all the computers then in existence? Supercomputers of the future, both classical and quantum in design, will have almost unimaginable potential and be augmented by equally astonishing developments in areas such as immersive technology and AI. These advances may help us find the means to reverse global warming, and overcome other grave threats that we face. On the other hand, ever more powerful and intelligent machines could very easily, and with alarming speed, become an existential threat in themselves.

CHAPTER 35

Reaching for the Sky

THE TALLEST STRUCTURE in ancient times is also the oldest of the Seven Wonders of the World and the only one to survive largely intact. Built around 2600 BCE, the Great Pyramid of Giza originally rose 146.5 metres (481 feet) from the desert floor and remained unsurpassed in height for the next 3,800 years. Only with the completion, in 1311, of Lincoln Cathedral, whose central spire climbed to 160 metres, was it overtaken.

Lincoln's spire collapsed in 1548, leaving a succession of churches in Germany and France as the tallest buildings on Earth between the sixteenth and nineteenth centuries. Nothing exceeded the old Lincoln spire in height, though, until the erection of the Washington Monument in 1884. This great obelisk-shaped building, which, at 169 metres, remains the tallest monumental column in the world, was overtaken after just five years by the Eiffel Tower in Paris.

Built as a centrepiece of the 1889 Exposition Universelle, which celebrated the start of the French Revolution, the wrought-iron, latticework tower was designed by engineers working for Gustave Eiffel's company. During its construction it became the first human-made structure to go past 200 metres, then 300 metres in height, finally topping out at 330 metres (1,083 feet). Not everyone had welcomed it. After its design was announced, 300 prominent Parisian artists and intellectuals signed a manifesto denouncing it as a 'gigantic black factory chimney'.

The original plan was to demolish the tower shortly after the Exposition finished. But since Eiffel had paid most of the bill for its construction, it was agreed to let it stand for twenty years so that he could recoup his investment. Eiffel realised he had to demonstrate the structure's usefulness other than as a controversial tourist attraction so he built an antenna on top and financed early experiments in wireless telegraphy. The tower proved so effective for sending and receiving wireless messages over long distances, particularly for the French military, that Eiffel's concession was renewed when it expired in 1909.

On the third and highest floor of the Tower, Eiffel installed a laboratory that he and French scientists used to study astronomy, meteorology and physiology. In 1909, he set up a wind tunnel at the base of the edifice that carried out thousands of aerodynamic tests, including those on the Wright brothers' planes and Porsche automobiles.

Thanks to strict safety measures only one person died during the tower's construction, though a few others have perished since. In 1912, a French tailor called Franz Reichelt jumped from the first floor wearing a homemade, spring-loaded parachute suit, which he'd hoped would eventually be used by aviators if they had to bail out at low altitude. He'd convinced the Parisian police to let him test the suit by claiming that he would use a manikin. However, on the fateful day, he showed up wearing the outfit himself, confident that he would glide to a safe landing. Sadly, the parachute only partially opened before almost immediately wrapping itself around his body. He plummeted like a stone, 57 metres to the frozen ground beneath the tower, creating a shallow crater.

Fourteen years after Reichelt's demise, military pilot Lieutenant Léon Collot made a bet that he could fly his plane through the arches at the base of the tower. He successfully negotiated the arches but then clipped a radio aerial as he made a turn and crashed into the Champs de Mars in a ball of flame. Although he survived the impact, he burned to death in the fire that engulfed his cockpit.

With the completion of the pinnacle of the Chrysler Building in New York in 1930 the Eiffel Tower was overtaken as the world's tallest structure. But the NY skyscraper's reign didn't last long. The following year, another art deco colossus, the Empire State Building, soared to new heights: 380 metres (1,250 feet) to its rooftop and 1,445 feet to the tip of its aerial.[1]

Today, the tallest building in New York is One World Trade Center, erected in the wake of the 9/11 attacks which destroyed the twin towers of the old World Trade Center. The architectural height (including a 124-metre spire) of 'Freedom Tower', as it's colloquially known, is 541 metres (1,776 feet). Its height in feet is a deliberate reference to the year in which the US Declaration of Independence was signed, and both the rooftop height and the ground footprint are the same as those of each of its fallen predecessors.

One World Trade Center, completed in 2016, is the tallest building in the United States and the whole of the Western Hemisphere. But it has been surpassed in height by several other skyscrapers in the East. Tallest of them all, by a wide margin, is the Burj Khalifa in Dubai, United Arab Emirates. It stands 828 metres – just over half a mile – tall to the tip of its spire. That's three times the height of the Eiffel Tower and almost twice as high as the Empire State Building.

The cloud-piercing Burj Khalifa, the top of which can be seen from 95 kilometres away, has 160 storeys. A public elevator runs 140 of those floors and, reaching 10 metres per second, can carry passengers from ground level to the observation deck on the 124th floor in just a minute.

No other building in the world has yet exceeded 700 metres, let alone 800 metres in height. Only two, aside from the Burj, currently stand taller than 600 metres: the Shanghai Tower in China and the Makkah Royal Clock Tower in Mecca, Saudi Arabia – the latter part of an enormous hotel complex intended, primarily, to cater for visitors and pilgrims to Islam's holiest site, the Great Mosque of Mecca, which stands nearby.

No new projects are underway, at present, which will overtake the Burj Khalifa. However, if work ever resumes on the Jeddah Tower, in Saudi Arabia, it would eventually claim the number one spot and, with a planned height of more than 3,280 feet, be the first 1-kilometre-tall building. About a third of the Tower has been completed but labour issues have stalled any further progress since 2016.

There's been no shortage of visionary plans for extraordinarily lofty buildings and other structures. Several of these have been proposed for Tokyo, the most populous city on Earth, with more than 37 million inhabitants in its metropolitan area. The Sky Mile Tower, plans for which were presented in 2015, would be built to a height of 1,700 metres (5,577 feet) on an archipelago of reclaimed land in Tokyo Bay.

Two decades earlier, an even more ambitious scheme for Tokyo Bay had been presented. The Shimizu City Mega-Pyramid would rise to fourteen times the height of the Great Pyramid of Giza and house 750,000 people. Not to be outdone, the X-Seed 4000 mega-tall skyscraper, designed by the Taisei Corporation in 1995, had a planned height of 3,778 metres – 224 metres taller than Mount Fuji, whose shape it echoes. X-Seed 4000, like all buildings that stretch upward several kilometres, would need to protect its occupants from changes in internal and external pressure and also weather fluctuations. In its case, the designers proposed using solar power to maintain a reasonably constant internal environment.

Another Tokyo-located structure, proposed in 1992, is the tallest building ever fully envisioned. The Tokyo Tower of Babel would ascend to a breathtaking height of 10 kilometres – surpassing Mount Everest as the highest point on Earth's surface. It would take more than a century to build, cost $22 trillion and house roughly 30 million people. All three of these ambitious schemes for the Japanese capital are examples of what's called arcology. The term was coined in the 1960s by Italian-born American architect Paolo Soleri from a combination of 'architecture' and 'ecology'. Arcology is a concept that's finding its way into many new large-scale designs, which combine space for various residential and commercial facilities with an attempt to minimise environmental impact.

Tall structures, in the sense of anything that's human-made and extends upward from the ground, aren't restricted to buildings. TV and radio masts need to be tall for obvious reasons and, for many years, were the loftiest structures anywhere. Nothing stood taller than the KVLY-TV mast in Blanchard, North Dakota, when completed in 1963, with a height of 629 metres (2,063 feet). It was surpassed eleven years later by the 680-metre-high Warsaw radio mast but this collapsed in 1991. Once again, North Dakota was home to the tallest structure in the world until the Burj Khalifa surged past it in 2008.

Only one other structure, of a sort, could greatly surpass in height any building currently standing or ever envisioned. The idea of a space elevator – a means of lifting loads into space using a physical connection that's anchored to the surface – was first described by Konstantin Tsiolkovsky in 1895. It has featured in many science fiction tales since and has been studied as a practical means of getting people and other payloads into orbit. Today's version of the space elevator is the space *tether*, conceived in 1960 by another Russian, Yuri Artsutanov.

Imagine a cable reaching from the ground, 35,800 kilometres vertically up into the black vacuum of space. At that height, a satellite completes an orbit in exactly the same time as it takes Earth to rotate once on its axis, so the satellite appears to hang at the same point in the sky. If the cable is fixed to an object in geosynchronous orbit, its ends don't move and it can be used as a means of transportation between the ground and space. The technological problems of building a space tether this long are immense. Chief among them: the tension in the cable would be so great that no known material would be strong enough and, at the same time, sufficiently lightweight to do the job. Perhaps in the future, breakthroughs in carbon nanotubes or diamond nanothreads will lead to a practical design, at which point the height record for human-made structures will be well and truly shattered.[2]

CHAPTER 36

Mesmerising Mechanisms

IN 1900, THE wreck of a Roman ship was discovered by sponge divers off the coast of the Greek island of Antikythera. Among the artefacts recovered from the vessel were the corroded remains of a device that is now widely believed to be the world's oldest calculator. For something made more than 2,000 years ago, it's astonishingly advanced in design and the intricacy of its parts.

The Antikythera mechanism, as it's become known, is housed in a wooden frame, measuring 34 by 18 by 9 centimetres, and consists of a series of interlocking bronze gear wheels, the largest of which has a diameter of 13 centimetres and, originally, 223 teeth. X-ray analysis of the device revealed a total of 37 interlocking gears, the movements of which, it seems, were used to calculate the movements of the Sun and Moon, predict eclipses and model the varying speed of the Moon in the sky (it being higher when the Moon is closer to us than when further away).[1]

Doubtless other such machines were built by Greek scientists between the third and first centuries BCE, but it wasn't until the fourteenth century in Western Europe that devices of similar complexity appeared again. In England, Richard of Wallingford, and in Italy, Giovanni de' Dondi, both made astronomical clocks that pushed the boundaries of what the technology of the day was capable. Dondi's Astrarium combined the functions of a clock, a calendar and a planetarium. Its 107 wheels and pinions, working in perfect symphony, could not only keep accurate track of the time and date but also calculate the motions of the Sun, Moon and five then-known planets, Mercury, Venus, Mars, Jupiter and Saturn.

Other elaborate and skilfully made mechanisms of the past were assembled for less practical purposes. In ancient Greece, the island of Rhodes had a reputation as a centre of mechanical engineering and was renowned for its playful automata. The poet Pindar, in the fifth century BCE, wrote: 'The animated figures stand adorning every public street and seem to breathe in stone, or move their marble feet.'

Over time, automata became ever more elaborate and lifelike. The twelfth-century Muslim scholar and inventor Ismail al-Jazari is sometimes referred to as the father of robotics for his surprisingly advanced creations. In his *Book of Knowledge of Ingenious Mechanical Devices* in 1206 he describes what are effectively programmable humanoid automata. His most remarkable contrivance was a boat that floated on a lake with four mechanical musicians aboard to amuse guests at drinking parties. The tireless band members were capable of dozens of different facial and body movements during their performances. Meanwhile, the percussion came courtesy of one of the earliest drum machines, operated by a series of pegs and tiny levers that could be programmed in advance to play a range of different patterns and rhythms.

Sometimes designs for ingenious and complicated devices ran ahead of the engineering techniques needed to build them. This was the case with the early mechanical computers invented by British mathematician Charles Babbage. He'd noticed that astronomical and other mathematical tables of the early nineteenth century were riddled with mistakes because all the calculations had to be done by hand and so were subject to human error. This gave him the idea of building a machine that would do the tedious work of computation more accurately, faster and without ever getting tired.

In 1822, Babbage was given a grant by the British government to build an automatic calculator, made from numerous interlocking gear teeth and connecting rods, which he called the Difference Engine. Construction started but never finished. Despite heroic efforts to construct a working model, the critical tolerances were beyond what engineers of the time could provide.

The government had spent £17,000, and Babbage £6,000 of his own money, on the project when Babbage set his sights on something even more ambitious. He grasped that the basic mechanisms of the Difference Engine could be generalised to an all-purpose calculating machine, programmable by a punched-card mechanism like that of a Jacquard loom. This vastly more powerful machine was called the Analytical Engine and would have been the world's first true computer. But it never got off the ground. 'He was ill-judged enough,' wrote the secretary of the Royal Astronomical Society, 'to press the consideration of this new machine upon the members of Government, who were already sick of the old one.' Prime Minister Robert Peel was less than enthusiastic: 'I would like a little previous consideration before I move in a thin house of country gentlemen a large vote for the creation of a wooden man to calculate tables from the formula $x^2 + x + 41$.'

The government's eventual withdrawal of support for his schemes left Babbage a disappointed and embittered man. However, his ideas survived and proved to be the forerunner of modern computers. Parts of his uncompleted mechanisms are on display in the London Science Museum. In 1991, working from Babbage's original plans, a Difference Engine was completed – and functioned perfectly.[2]

The first true computers emerged out of studies carried out, at both the theoretical and practical level, in the 1930s. At the Massachusetts Institute of Technology, in Cambridge, MA, Vannevar Bush developed the first modern analogue computer called the Differential Analyzer. Its moving shafts and gears handled long series of additions and multiplications, while a knife-edged wheel rotating on a circular table was involved in integration – a fundamental process for solving many problems in science and engineering. The Differential Analyzer, busily and noisily at work, can be seen early on in two popular science fiction films from the 1950s: *When Worlds Collide* (1951) and *Earth vs. the Flying Saucers* (1956).

On the other side of town from MIT, and around the same time, Howard Aitken at Harvard University was working on a different approach to computation – *digital* rather than analogue. Starting in 1937, he designed a series of four calculating machines of increasing sophistication, from the mostly mechanical Mark I to the fully electronic Mark IV. By the end of the Second World War, vacuum tubes, or valves, had replaced mechanical gadgetry and the age of computers – those most extraordinary of human inventions – had begun.

Vacuum tubes, though, were notoriously unreliable and prone to burning out. When thousands of them had to work in concert in the early electronic computers of the 1940s and 1950s, it was a relentless job to replace the ones that had broken down. The invention of the transistor in 1948 and its subsequent dramatic miniaturisation revolutionised

computing and the world that became increasingly depend-
ent upon it.

Fascination with making things smaller and smaller has
been a recurrent theme in science fiction for a century or
more. Sometimes people are shrunk accidentally by radiation
or, as in the TV series *Land of the Giants*, by a mysterious
disturbance in spacetime. In other cases, the miniaturisation
is brought about intentionally by some breakthrough in technology. The 1966 film *Fantastic Voyage* is about a submarine
and its crew that are reduced to the size of a microbe and
then injected into a patient in order to clear a blood clot in
his brain.

No such means of making things smaller by compressing
the original, down to the atomic scale, is feasible. Dramatic
strides in miniaturisation have been possible, though, by
finding new, smaller components. First the transistor, then
the integrated circuit, in which many, microscopic transistors
could be fabricated on a single silicon chip, allowed computers
to shrink in size and become ever more powerful. Today's
most advanced semiconductor chips contain many billions
of transistors in an area not much larger than a thumbnail.
The current record is 2.6 trillion transistors – each a separate
switch that can be used to process or store data – held by
Cerebras's Wafer Scale Engine.

At the cutting edge of technology today are molecular
machines. As the name suggests, these are devices consisting
of individual molecules, or assemblages of molecules, that
are tailored to perform simple tasks. Natural examples exist
in the form of ribosomes, which help assemble proteins, and
kinesins, which transport other molecules from one place to
another. Artificial molecular machines (AMMs) are still in the
early stages of development but already a variety of molecular motors, switches and logic gates has been demonstrated.

Microscopic submarines, like that in *Fantastic Voyage*, may always be a pipedream but 'nanobots', able to roam around inside our bodies, delivering drugs with pinpoint precision, or hunting down and destroying cancer cells, are very much on the horizon.

CHAPTER 37

Whoosh

THE FASTEST HUMAN on record is the Jamaican sprinter Usain Bolt who, in 2009, averaged 37.7 kilometres per hour in covering 100 metres in 9.58 seconds. His top speed, reached in the 60–80-metre stretch, was 44.7 kilometres per hour. Impressive as that is, Bolt would have been left for dust by the fastest of all two-legged animals, the ostrich, which can sprint at up to 70 kilometres per hour and, more remarkably, keep up a steady pace of about 60 kilometres per hour for thirty minutes at a stretch.

Four legs, though, are better than two in the speed stakes. Nothing is faster on land than the cheetah, whose light build, long thin legs and long tail allow it to travel at 98 kilometres per hour or more in short bursts. This is only slightly more than some of its prey, including the pronghorn and the springbok, but the cheetah has the additional advantage of exceptional acceleration and manoeuvrability.

In water, nothing outpaces the streamlined sailfish, which can reach 110 kilometres per hour. But even this is made to seem pedestrian by the fastest animals on Earth. Birds take all the speedster top spots. When it comes to level, flapping flight the record is contested by several birds that can fly in the 150–170-kilometres-per-hour range: the spine-tailed swift, the Eurasian hobby and the frigatebird, the last being helped by having the largest wing-area-to-body-weight ratio of any bird. When dives are factored in, a clear winner emerges in the speed stakes: the peregrine falcon, which in its hunting dive, or stoop, can travel at an astonishing 389 kilometres per hour (242 miles per hour).

Throughout most of human history, the quickest way to get from one place to another was by horse. But although horses are fast they can't gallop for very long or they overheat. The solution is to switch horses at regular intervals. Such relays date back nearly 4,000 years and were widely used in ancient Babylonia, Persia, China, Mongolia and Egypt. One of the most famous, in more recent times, was the Pony Express of the American West, which offered the swiftest method of getting a letter between the Atlantic and Pacific coasts before the first railroads spanned the country.

A horse-drawn stagecoach took twenty-five days or more to make the transcontinental journey – three weeks faster than sailing around Cape Horn but still pretty slow. With the Pony Express, railroads handled the eastern leg of the journey before horsemen, racing day and night between relay stations 16 kilometres apart, completed the trip from Missouri to California. The Pony Express could get a small bag of mail across the width of the US in ten days for just a few dollars per letter. But it operated for just eighteen months, between April 1860 and October 1861, before closing down a few days after cross-country telegraph became available.

The first vehicles capable of travelling faster than horse were early trains. The Liverpool and Manchester Railway, which ushered in the era of passenger trains, ran at up to 48 kilometres per hour. But by 1850, steam locomotives were pulling British trains at up to 125 kilometres an hour – faster than any animal on land or sea. In 1938, the powerful, streamlined *Mallard* set a world speed record for a steam locomotive of 203 kilometres per hour, which stands to this day.

The first cars were slower than horse-drawn carriages and, in Britain, restricted by what became known as the Red Flag Act of 1865. This made it illegal to drive at more than 4 mph on country roads and 2 mph in towns, and further required all self-propelled vehicles to be preceded by a person walking at least 60 yards ahead carrying a red flag. There was a hefty fine of £10 for 'speeding'. The act was finally repealed in 1896 when the speed limit was raised to 14 mph.

Until 1903, trains held the land speed record for vehicles in which people could travel. In that year the accolade went for the first time to a car – a Gobron Brillié was clocked at 132 kilometres per hour. In 1906, the Stanley *Rocket*, an American steam-powered automobile, became the first vehicle to travel at more than 200 kilometres per hour and the fastest steam-powered land vehicle of all until 2009.

By the early 1920s, humans had a new quickest way to travel – by air, in planes that could surpass 300 kilometres per hour. Never again would any form of transport on land or sea be able to outpace the fastest of aircraft. By 1923, the air speed record stood at more than 400 kilometres per hour. At the outbreak of the Second War World, two German planes had surpassed 700 kilometres per hour. One of these, the Heinkel He 100, was the fastest fighter plane in the world at the time of its development but for unknown reasons was never ordered into production. Only nineteen prototypes and pre-production examples were built and none survived the war. The other pre-war German speedster, the Messerschmitt 209, was a single-engine racing aircraft designed for breaking speed records and to serve as a prop-aganda tool.

During the war, three planes manufactured by Messerschmitt became the first to fly at more than 1,000 kilometres per hour. Two of them, different versions of the Me 163 Komet, were the only operational rocket-powered fighter aircraft in history and the first piloted aircraft of any kind to exceed 1,000 kilometres per hour in level flight.[1] The Me 262, the first operational jet-powered aircraft, attained a similar top speed but only during a steep dive.

In the immediate post-war years, a race began to be the first to travel faster than the speed of sound – to break the sound barrier. In the UK, progress was made towards this goal in the form of the Miles M.52 turbojet-powered aircraft. Following an agreement between the British Air Ministry and the United States, details of the UK's high-speed research and designs were shared with the US. Bell Aircraft was given access to the plans for the M.52 and used these to initiate work on the Bell X-1, the final version of which had many similarities with the M.52.

On October 14, 1947, Air Force Captain Charles 'Chuck' Yeager took his rocket-powered Bell X-1 to a speed of 1,299 kilometres per hour, or Mach 1.06 – 6 per cent faster than the speed of sound.[2] The following month he shattered that record in the same plane by reaching 1,434 kilometres per hour (Mach 1.17). There was fierce inter-service rivalry in those days to push the boundaries of airspeed. On November 20, 1953, the Navy's rocket- and jet-powered D-558-II Skyrocket was flown by Scott Crossfield, a test pilot with the National Advisory Committee (NACA) High-Speed Flight Station at Edwards Air Force Base in California to a speed of 2,078 kilometres per hour, a shade over twice the speed of sound and 25 per cent more than the Skyrocket's intended design speed.

Having been comprehensively bested, Yeager and his fellow test pilot Jack Ridley were determined to reclaim the airspeed record for the Air Force in a series of flights they dubbed 'Operation NACA Weep'. On December 12, 1953, Yeager took off in a Bell X-1A, a plane similar to the X-1 but with turbo-driven fuel pumps, a new cockpit canopy, longer fuselage and increased fuel capacity. Having climbed to 80,000 feet, he accelerated to a world-beating Mach 2.44 before disaster nearly struck. The aircraft began to spin out of control along all three axes, the g-forces rose disastrously, reaching up to 8 g, and Yeager's head was thrown forward so violently that his helmet cracked the plane's plastic canopy. The X-1A spun down 51,000 feet in 51 seconds before, in the denser air at 25,000 feet, Yeager was able to regain control and land safely. Not only had Yeager and Ridley achieved their aim of breaking the Navy's record but they did it in time to spoil a celebration the rival service had planned for the fiftieth anniversary of powered human flight at which Crossfield was to have been named 'the fastest man alive'.

World record speeds on land and water and in the air have all been shattered since the 1950s but, surprisingly, no new ones have been set yet in the twenty-first century. The land speed record is currently held by Wing Commander Andy Green, a retired British fighter pilot who, in 1997, drove the jet-powered Thrust SSC to 1,223.7 kilometres per hour (760.3 miles per hour) and, in the process, became the first person to break the sound barrier without leaving the ground. The water speed record has proven to be the most hazardous to attempt. Of the thirteen people who have tried to break the record since 1930, seven have died as a result. The current record of 511 kilometres per hour is held by Australian motorboat racer Kim Warby, who achieved it in 1978 on Tumut River near the Blowering Dam in New South Wales with *Spirit of Australia*, a jet-powered, wooden speedboat built in a Sydney backyard.

As for the air speed record, that depends on what qualifies. The Lockheed SR-71 Blackbird is the champion among manned air-breathing jet aircraft, with a top speed of 3,550 kilometres per hour (Mach 3.2) achieved in 1976. Among rocket planes the record holder is NASA's North American X-15, which, in 1967, was flown by William 'Pete' Knight to 7,274 kilometres per hour (Mach 6.7) at an altitude of 31,120 metres (19.3 miles). No crewed, powered aircraft has ever travelled faster. The only human-carrying vehicles that have reached greater speeds are those lifted into space by rockets or that have re-entered the atmosphere, unpowered. On November 14, 1981, the Space Shuttle *Columbia*, piloted by Joe Engle, achieved the fastest manually controlled flight in the atmosphere during re-entry of Shuttle mission STS-2.

THE NATURAL WORLD

CHAPTER 38

The Many and the Few

As I write, there are, according to the online 'Worldometer' (based on UN data), just over 8 billion people on Earth. That's up from 2.5 billion in 1950 and 1.9 billion in 1920. So there are people alive today who were born when the world population was less than a quarter of what it is now.

China and India are the two most populous countries, each with about 1.4 billion inhabitants. It's estimated that, throughout human history, about 110 billion people have died, so that roughly 7 per cent of all the people who've ever lived are alive at present.

By 10,000 BCE, the world's population was around 1 million. By 1 CE, that number had jumped to 200 million – 1 million of which lived in the largest city at the time, Rome. Figuring out the human population in early times is tricky because there are no remains of large settlements and no census results or written records of any kind. Estimates have to be based on evidence such as that of bones, fossils and DNA sequencing.

There's a big difference between the total population of a species and the number of breeding individuals at any given time. Another important factor is the average life span, which was much lower in prehistoric times. Between 4 million and 200,000 years ago, when we and our hominid ancestors lived exclusively on the African savannah, the average lifespan was about twenty years, so that almost the entire hominid or human population would be completely renewed about five times every century. Our ancestors' global population during this long period, upon which the future of the human race depended, fluctuated between about 100,000 and as few as 10,000 individuals. In terms of conservation status, we and our forebears would probably have been classified as 'vulnerable' or 'endangered'.

Among larger mammals today, we rank among the most numerous. We certainly top the list when it comes to primates. Our nearest rival in this department is the long-tailed macaque which – surprise, surprise – has a long history of association with humans. There are thought to be about 2.5 million of these omnivorous monkeys in various parts of Southeast Asia. Though, like all non-human primates, they've suffered from habitat loss, they've also proved to be opportunistic and highly adaptable, stealing food from people, both passively and aggressively, and making themselves especially unpopular with farmers. No other living primate can boast a wild population of more than about a third of a million, and many species have populations that number in the few thousands or few hundreds.[1]

The only mammals that might outnumber us today are two animals that have benefited hugely from our presence – the brown rat and the house mouse. Both, of course, are considered pests and vermin, although, in the broader scheme of things, it would be hard to name a species that has had a more deleterious effect than *Homo sapiens*!

It's also no surprise that among other numerous mammals are those that we've domesticated, either as farm animals or pets. There are about 1.5 billion cows, the majority of them in India (where they're considered sacred), Brazil and the US. Dogs number around 470 million and pet cats about 370 million.

In the avian world, the commonest species in the wild is reckoned to be the red-billed quelea, native to Africa south of the Sahara. This sparrow-sized bird nests in enormous colonies. A single tree may be hung with hundreds or even thousands of carefully woven nests. Single colonies can cover hundreds of acres, totalling tens of millions of individuals. Altogether the population has been put at about 1.5 billion.

But the most numerous of all vertebrates are not land-dwellers but fish. Estimates put the piscine population of the world's oceans at about 3.5 trillion. The number of known fish species – about 5,600 – is split roughly evenly between saltwater and fresh. And the most common fish of any species is one that perhaps you've never heard of. Called a bristle mouth, it's the size of a small minnow and lives at depths below 500 metres all over the world.

It's generally the case that small things outnumber large, and so it is with living creatures. There are far more invertebrates than animals with backbones. Of the 2.5 million species of animal that have been documented, about 1 million, or 40 per cent, are insects. But researchers believe there are many more insect species – perhaps between 10 and 30 million – that have yet to be described by science.

There are roughly 400,000 species of beetle alone. That means there's more than one beetle *species* for every living individual chimpanzee in the wild. In fact, beetles (*Coleoptera*) account for about a sixth of all the animal species in existence. There are fewer species of ant (about 14,000) but in terms of individual numbers, ants dominate the insect world with somewhere between 10 trillion and 100,000 trillion individuals globally. It's quite possible that, combined, ants outweigh the entire human race.

Of course, there are huge numbers of living things on the planet that are too small for us to see without a microscope. Smallest, most primitive and most numerous of all are bacteria and archaea, collectively known as prokaryotes. The number of bacteria inhabiting your large intestine is a staggering 40 trillion, give or take a few trillion. Microorganisms make up between 1 and 3 per cent of our total body weight. Put another way, if you weigh, say, 70 kilograms, you're carrying around one or two kilos of bacteria. You're welcome!

The number of bacteria on Earth has been put at about 20 million trillion trillion trillion, or 2×10^{30}. Their biomass is exceeded only by that of plants. Even if all the higher forms of life were to be virtually wiped out by some natural cataclysm, such as a large asteroid strike, or human stupidity, there would still be bacteria and possibly some other hardy forms, such as tardigrades and cockroaches.

While, for the time being, humans, rats and other adaptable species continue to prosper, many animals are threatened with extinction. There are only about seventy-five Javan rhinos left in the world, all of them confined to the Ujung Kulon National Park, a World Heritage Site. The Gobi bear, a subspecies of brown bear found only in the Gobi Desert of Mongolia, is down to its last fifty-one individuals. The saola, also called the Asian unicorn, a forest-dwelling bovine native to a small region of Vietnam and Laos, was discovered as recently as 1993 and may number as few as twenty in the wild. Although many attempts have been made to breed them in captivity, they've all died within a matter of weeks or months.

The rate of extinction of animals and plants is increasing. The last sighting of many has been within living memory: the Ethiopian amphibious rat (1927), the Northern Sumatran rhinoceros (after 1960) and the baiji, a freshwater dolphin that once lived in the Yangtze River (2002). Several hundred species have become extinct over the past decade and the rate of loss is accelerating. We're merely at the start of the sixth great mass extinction the world has known – and the first caused by human activity.

There's hope that some species might be brought back to life – a process sometimes referred to as 'de-extinction' – through the use of techniques such as cloning or selective breeding. However, concerns have been raised over whether this is a good idea. Reintroducing an extinct species could, for instance, have a negative impact on extant species and their ecosystem. Only a drastic change in the relationship between *Homo sapiens* and the biosphere upon which we depend, it seems, is likely to avert a catastrophic loss of biodiversity.

CHAPTER 39

Descent

In Jules Verne's *A Journey to the Centre of the Earth*, Otto Lidenbrock, an eccentric German scientist, his nephew Axel and their Icelandic guide Hans descend into the crater of the volcano Snaefellsjökull. Far underground they encounter an ocean and creatures still alive from the age of the dinosaurs. Reality, it turns out, isn't far behind. A giant reservoir of water *has* been detected at spectacular depths. Strange life-forms *have* been found that never see the light of day. And people have found ways to penetrate deeper and deeper into our planet's interior.

Natural caves have served as shelters and dwellings since the dawn of our species and, in historical times, were the source of myth and rumour. The Alepotrypa Cave in the Peloponnese peninsula is one of several Greek locations linked with the subterranean domain of Hades. Belief in an underground realm of lost souls may have evaporated, but conditions deep below the surface can certainly be hellish in a physical sense.

People have been exploring caves for millennia but the *science* of caving – speleology – began less than 200 years ago. In the first half of the nineteenth century, our knowledge of karst – topography sculpted by the action of rainwater on limestone and other soluble rocks – grew dramatically through the exploration of caves around Postojna, in Slovenia. Near Trieste, the Abisso di Trebiciano was explored in 1840 to a depth of 320 metres below the entrance and for the next sixty years was the deepest known cave in the world.

Today, the record depth, of 2,212 metres (7,257 feet), is held by the Veryovkina Cave in the Abkhazia region of Georgia.[1] The entrance was discovered in 1968 by Soviet speleologists at a height of 2,285 metres above sea level in the Gagra mountain range of the West Caucasus. All four of the currently known deepest caves are in the same locality.

Karst covers up to a quarter of Earth's land surface and much of it is riddled with passages eroded in the rock by mildly acidic water. There may be tens of thousands of caves awaiting discovery – some of them, almost certainly, extending deeper than Veryovkina. The only limit may be how far groundwater can penetrate into the limestone before the pressure becomes too great for it to flow.

No such limits apply to artificially made caves. All of the twenty deepest mines ever excavated exceed the Veryovkina cave in depth. At the top of the list are half a dozen South African gold mines, headed by the Mponeng mine in Gauteng province. The trip, in stages, from the surface to the lowest point, 4 kilometres (2.5 miles) down – ten times the height of the Empire State Building – takes over an hour. The temperature of the rock at the greatest depth reaches an insufferable 66 °C and to cool the tunnel air to a tolerable 30 °C, salty ice-slurry is pumped underground.

It might seem as if the heat and crushing pressure of the rock would create an insurmountable barrier to life. But, in 2008, researchers found specimens of the bacterium *Desulforudis audaxviator* present within groundwater near the base of the Mponeng mine. The name of the microbe comes from a quote in *A Journey to the Centre of the Earth*, where Professor Lidenbrock finds a secret inscription in Latin: *Descende, audax viator, et terrestre centrum attinges* (Descend, bold traveller, and you will attain the centre of the Earth).

What of places that are low down without being underground? The lowest point on dry land is the shore of the Dead Sea at 433 metres below sea level. But this is easily surpassed by much of the land that lies below a thick layer of ice in Antarctica. Under the Denman Glacier, the bedrock starts at a depth of 3,500 metres below sea level – not much less than the distance to the bottom of the Mponeng mine. For the lowest point anywhere on Earth's surface, we must look to the Challenger Deep, 11,034 metres below sea level, in the Mariana Trench. Only twenty-seven humans have ventured to this extraordinary place, including Jacques Piccard and US Navy Lieutenant Don Walsh in 1960 aboard the bathyscaphe *Trieste*, filmmaker James Cameron in 2012 aboard *Deepsea Challenger*, and Victor Vescovo, Patrick Lahey and Jonathan Struwe aboard the *DSV Limiting Factor* in 2019.

You might suppose that the Challenger Deep is the closest that humans have come to the centre of the Earth. But that would ignore the fact that our planet isn't exactly round. In fact the closest point anyone can get to the centre is at the bottom of Litke Deep, off the coast of Greenland in the Arctic Ocean. It lies a mere 6,351.7 kilometres from the centre – 14.7 kilometres closer than the Challenger Deep.

Fortunately, we don't have to descend in person to ever-greater depths to learn more about our planet's internal makeup. Seismic waves – shock waves produced by earthquakes and explosions – travel through the Earth and across its surface, acting like a geological version of X-rays. They've revealed a layered infrastructure, with a solid rocky crust overlying a much thicker, semi-solid mantle, consisting of upper and lower regions, which, in turn, sits atop the core, again divided into outer and inner parts.

The crust varies in thickness from between 30 to 50 kilometres beneath the continents and a bare 5 to 10 kilometres under the oceans. Between the crust and mantle is a sharp transition zone called the *Mohorovičić* discontinuity (or 'Moho'). The crust and top part of the mantle, known collectively as the lithosphere, is made mostly of solid rock, but at the boundaries between tectonic plates the mantle material is more fluid, enabling the continents to drift and slide relative to one another.

Scientists would love to get their hands on some fresh mantle material, especially the rocks lying on either side of the Moho. But this means first penetrating through the crust, to depths not reached by any cave or mine. The only feasible way of bringing back samples from the mantle is by drilling a hole, straight down, until it passes through the crust–mantle boundary.

The first attempt to drill through to the mantle was Project Mohole, conducted in the 1960s by US scientists.[2] The Soviet Union was rumoured to be considering a similar project so, not surprisingly, the media was quick to portray Mohole as the American contribution to a terrestrial version of the Space Race. It began by sailing a repurposed oil ship, the *CUSS I*, to a location near Guadalupe Island off the coast of Mexico. The vessel was able to maintain its position accurately by a new technique that involved four large outboard motors on the ship, positioning the vessel within surrounding moorings using acoustic techniques, and guiding the motors by a central joystick.

Drilling down into the seabed made sense because, as mentioned, the oceanic crust is several times thinner than the continental crust. On the other hand, the research team had to figure out how to lower long segments of pipe, which were constantly buffeted by strong ocean currents, and send the drill down through the pipes before boring through the crust could even start. There was also the problem of how to bring up core samples of rock and mud, from great depths, in the form of intact cylinders.

The Mohole scientists opted for a cautious approach – first drilling down a short way into the crust to make sure the technology would work. Five holes were drilled, the deepest to 183 metres below the seafloor in 3,600 metres of water. The core samples proved valuable, confirming that the properties of the sediments and rocks recovered matched the findings of earlier seismic studies. This first phase of the drilling was heralded as a great success by both the scientific community and the oil industry. But then disaster struck. The various academic bodies involved in the programme couldn't agree on what the next steps should be; criticism was voiced about the cost of the project, and, as usual, there was political bickering. In the end, Congress refused further funding and Mohole died – but not before it had shown that the means were available to drill into Earth's mantle.

In 1970, the Soviet Union began its own attempt to break through the crust on the Kola Peninsula, near the Russian border with Norway.[3] Nearly two decades later, the Kola Superdeep Borehole project reached a depth of 12,282 metres. This is currently the deepest artificial point on the planet, although because the project took place on land, it's only about a third of the way through the continental crust, which is estimated to be about 35 kilometres thick at the drilling site. Further progress was hampered by equipment breakdowns and the fact that the rock, at 180 °C, was much hotter than expected, causing it to behave like a plastic and making drilling almost impossible.

Today, the quest to break through the crust and obtain a pristine chunk of mantle continues. And the prize is great: the mantle makes up about 68 per cent of our planet's mass and an impressive 85 per cent of its volume. Being able to analyse its composition, unaltered by mixture with crustal rock and other contaminants, would help scientists learn more about the raw materials from which Earth accreted in the early days of the Solar System.

Between 2002 and 2011, four holes at a site in the eastern Pacific managed to reach fine-grained, brittle rock that geologists believe to be cooled magma sitting just above the Moho. But the drill couldn't punch through those last tenacious layers. In 2013, drillers at the nearby Hess Deep found themselves similarly limited by tough deep-crustal rocks. Most recently, in 2019, the Japanese ship *Chikyu* drilled to a record 3,250 metres beneath the seafloor.

We haven't yet been able to follow all the way in the footsteps of Lidenbrock's party. But volcanoes, as in Verne's tale, have given us access to some of the secrets hidden deep inside the Earth. Every time they erupt they bring material, originally from the mantle, which has mixed with crustal rocks inside magma chambers, to the surface. Perhaps within the next decade or so, researchers will have some pure mantle specimens to examine in the lab.

CHAPTER 40

Age Matters

MODERN HUMANS EMERGED in Africa about 300,000 years ago – within the most recent 0.002 per cent of the lifetime of the universe. The earliest settlements date to around 10,000 BCE and the first civilisations to a mere 4,000 BCE. It's hard to imagine, in these days of instant communications and the World Wide Web, that there are no written records extending back more than about 6,000 years.

Before they had a written or even a complex spoken language, our ancestors made things, often very skilfully, for various purposes, including art and music. The 'Venus of Hohle Fels' is a figurine fashioned of woolly mammoth ivory that was found in the Hohle Fels Cave in Schelklingen, Germany, in 2008. Dated at between 35,000 and 40,000 years old, it's a female figure with grossly exaggerated anatomy that may have served as a symbol of fertility. It's the oldest depiction of a human form known. Similar in age is the 'Lion-man of Hohlenstein-Stadel', again carved in ivory, which, though generally anthropomorphic, has the head of a European cave lion.

Music has its roots long before people lived in towns and villages. Flutes, made at least 42,000 years ago from bird bone and mammoth ivory, were discovered at Geissenkloesterle Cave in southern Germany. From the positioning of the holes, it's clear that they were able to play distinct melodies and may have been used for religious rituals or simply for recreation. Older still is the 'Neanderthal flute', from the cave of Divje Babe in Slovenia, carved from the bone of a cave bear 50,000 years ago.[1] Remarkably, its four finger holes enable it to produce notes that exactly match the diatonic scale used in music today.

Art, too, flourished tens of millennia ago. In the Blombos Cave of Western Cape, South Africa, were found two abalone shells containing traces of a red paint-like mixture. Nearby, researchers came across ochre (coloured clay), bone, charcoal, stone hammers and grindstones, which, it's thought, were used by our palaeolithic ancestors to manufacture paints. The oldest known painting is a life-sized picture of a wild pig in the Leang Tedongnge Cave in a remote valley in Indonesia. It was created using dark red ochre pigment. On top of some of this pigment, however, has been deposited some calcite that's been accurately dated at 45,500 years. So the painting must be at least this old and possibly very much older.

Most ancient of all hominid artefacts are stone tools, used for such tasks as cutting, chopping and scraping. The Lomekwi stone tools, from West Turkana (formerly known as Lake Rudolf) in Kenya, predate any others yet found. They were manufactured about 3.3 million years ago by a species that lived on the African savannah before any member of our own genus, *Homo*, had emerged.

If the whole history of Earth, since its formation just over 4.5 billion years ago, is represented by a single year, our earliest hominid ancestors first diverged from the line that led to modern apes, about 14.5 hours ago. Anatomically modern humans appeared within the past half-hour. We're newcomers to a world that's ancient and a universe that is three times more ancient still.

The oldest known minerals on Earth are zircon crystals that were found inside rock from the Jack Hills region of Western Australia.[2] They formed 4.4 billion years ago, not long after the surface of our planet solidified from its initial molten state. But there are materials of even greater age, which have been found on Earth but that came from elsewhere.

Recently, while running an exhibit at the Dundee Science Centre, I was approached by a man who lives in Winchcombe, Gloucestershire. On the night of February 28, 2021, this quiet market town on the edge of the Cotswolds became the centre of a national, not to say international, news story. A meteorite had been seen to fall around Winchcombe. It broke into fragments and one piece had landed on the driveway of Rob Wilcock, making a small crater. My visitor at the Science Centre was a neighbour of the Wilcocks and a volunteer at the town's museum, where three small bits of the meteorite now reside. Witnessed meteorite falls are very unusual, and this particular meteorite turned out to be of a rare type called a carbonaceous chondrite. Analysis revealed that it had formed about 4.6 billion years ago from primordial matter going around the infant Sun, at a time when the planets themselves were still coalescing from the solar nebula.

Among other stars and planets are many that existed long before our own Solar System took shape. When the universe was new, everything in it consisted almost entirely of two elements – the two lightest ones, hydrogen and helium. For this reason, any stars that have survived to the present day from those early times ought to be 'metal poor'. In astronomy, all elements heavier than hydrogen and helium are referred to as 'metals' even though this includes elements such as carbon and oxygen, which are definitely not metals in the normal sense.

Some 190 light-years away, in the constellation Ophiuchus, lies HD 140283, known popularly as the Methuselah star.

A study in 2013 suggested that its age is 14.5 billion years plus or minus 0.8 billion years. The problem with that is that the universe itself is thought to have begun about 13.8 billion years ago. How could a star be older than the universe? More recent theoretical work in stellar modelling has revised Methuselah's age to no more than 13.7 billion years – extraordinarily great but at least no longer an astrophysical embarrassment.[3]

HD 140283 and other ultra-metal-poor stars like it are thought to be second-generation stars. The first stars in the universe, composed exclusively of hydrogen and helium, were probably very big and massive and raced through their brief lives before exploding as supernovae. The debris from these early blowouts would have included a smattering of heavier elements, formed inside the first stars before they self-destructed, which explains why second-generation stars such as Methuselah do contain small amounts of 'metals'. One of the hopes for the James Webb Space Telescope, launched in December 2021, is that it will give astronomers a glimpse of the very first stars to shine after the dawn of time.

When it comes to matters of age, nothing is older than the event in which the universe began – the Big Bang. The temperatures within the first second or two after that primordial eruption were inconceivably high, as was the energy of the radiation pouring out of it. But the universe is now about 13.8 billion years old and the remnants of that long-ago fireball have cooled through aeons of cosmic expansion. Another way to think of this is that the stretching of spacetime, from the moment of creation to the present, has enormously increased the wavelength of the radiation that came out of the Big Bang.

Today, the genesis glow, coming from every direction in the sky, has been shifted into the microwave region of the spectrum, corresponding to a temperature of just 2.7 degrees above absolute zero. This is the cosmic microwave background, discovered in the 1960s, and the oldest phenomenon we've ever encountered – and, perhaps, ever will.

CHAPTER 41

Small

LIFE IS PERILOUS if you're an Etruscan shrew. Being the smallest mammal in existence – 4 centimetres long and the weight of a playing card – makes you easy prey. The main problem for these tiny insectivores, though, is their exceptionally high rate of metabolism, which means they need to eat up to eight times their own body weight every day. They can't hibernate because they'd starve to death. So, instead, they have a neat trick for cutting back on their energy use in winter: they shrink their brains. They lose more than a quarter of the neurons in the sensory part of the brain responsible for processing information from their whiskers, then regrow these neurons the following spring.[1]

The world's smallest bird, the bee hummingbird, a native of Cuba, also lives life in the fast lane. Often mistaken for the insect after which it's named, it beats its wings about 80 times a second, although males can step this up to 200 times a second during their courtship flight. Bee humming-birds weigh less than 2 grams and the female lays eggs the size of a coffee bean in a nest the width of an American quarter coin.

Among vertebrates – animals with backbones – the smallest of all is a frog the size of a housefly. *Paedophryne amauensis* grows to no more than 8 millimetres in length and was discovered as recently as 2009 by scientists on an expedition in Papua New Guinea.

Close to the limit of what we can make out with the unaided eye is the parasitic wasp *Dicopomorpha echmepterygis*, which belongs to a family of tiny insects commonly known as fairyflies. Males are blind, wingless and a mere eighth of a millimetre long. To see anything smaller, some kind of artificial aid is needed.

The first on record to employ a form of magnifying lens was the English friar and scholar Roger Bacon in 1268. He used glass spheres to create enlarged, albeit distorted images of things at the limit of human vision. The idea of using combinations of lenses to increase magnification was described by Girolamo Fracastoro of Verona in his book *Homocentricorum sive de Stellis* of 1535. Then, towards the end of the sixteenth century, the first compound microscopes started to appear, opening the path to investigations of the microscopic world. Robert Hooke's *Micrographia* (1665) contained wonderful, exquisitely drawn illustrations of insects and plants seen through this new optical instrument. Hooke famously gave the first detailed description of a fly's eye and a plant 'cell' – coining the term because of the resemblance of what he saw to the cells of a honeycomb.

A dozen years after Hooke's eye-opening pictures appeared, the Dutch microscopist Antonie van Leeuwenhoek wrote his famous 'letter on the protozoa' in which he announced his discovery of bacteria and protists – single-celled organisms that aren't plants, animals, bacteria or fungi. It was many years before others could match the resolution and clarity of his instruments, so his findings were at first doubted and even dismissed.

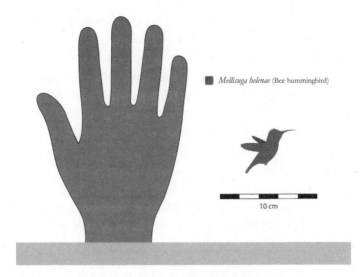

Mellisuga helenae (Bee hummingbird)

10 cm

Comparison in size between a hand and a bee
hummingbird, *Mellisuga helenae.*

Today, more than 30,000 species of bacteria have been identified, the smallest of which are just a few hundred nanometres
(billionths of a meter) in size. Much smaller still are viruses –
in some cases no more than 20 nanometres across – though
these aren't regarded as independently living organisms.

Descending further in size, and into a realm only electron microscopes can penetrate, we come to macromole-
cules, such as DNA and proteins, then to smaller molecules.
Finally, individual atoms come into view. The most powerful
microscopes on Earth can distinguish individual atoms, each
measuring between one tenth and five tenths of a nanometre
(0.1–0.5 nm) across.

At the heart of every atom is the nucleus, where almost all
the atomic mass, in the form of protons and neutrons resides.
Fantastically small though an atom may be, the diameter of

its nucleus is 10,000 times less. If the nucleus were the size of a blueberry, the atom would be roughly as big as a football stadium.

A proton is unimaginably tiny, a mere 1.7 femtometres across. A femtometre is one quadrillionth, or thousand trillionth, of a metre (10^{-15} m). Until 1964, it was thought that protons and neutrons, like electrons, were elementary – in other words, indivisible into anything smaller. But now we know that each contains three particles called quarks.

According to our best available theory of the subatomic world, known as the Standard Model, quarks, electrons and other truly elementary particles occupy mere points in space. In practice, we can't measure something that has no extent whatsoever. The best we can do is collide other particles with protons in an effort to probe the details of their constituents. Experiments carried out with the HERA collider in Germany – effectively, the world's most powerful electron microscope before it was shut down – set a new upper limit on the quark's diameter. It is, at most, 1/2,000th the size of a proton. In turn, the proton is about 60,000 times smaller across than a hydrogen atom, which is about 40 times smaller than the diameter of a DNA helix, which is about a millionth the size of a grain of sand.

Quarks, electrons and other elementary particles are dimensionless points in the Standard Model. But we know that the Standard Model can't be complete because there are vitally important questions about the universe that it leaves unanswered. How does gravity fit into the quantum scheme of things? Why is there so much matter and so little antimatter around today? And what on earth is dark matter and dark energy?

New physics is waiting to be found, and when it is we may learn that the likes of quarks and electrons have a tiny but non-zero size and that there are even more minute components of nature. Theories that go beyond the Standard Model, such as supersymmetry and string theory, predict finite sizes for quarks and electrons. In the so-called Rishon model, all quarks are composed of three, even smaller units called preons. In string theory, the fundamental building blocks of matter aren't zero-dimensional particles but rather, vibrating one-dimensional strings. How long is a piece of string? According to string theory, the average string, if it exists, is about 10^{-35} metres long. That's a hundredth of a millionth of a billionth of a billionth of a billionth of a metre. Magnify an atom to the size of the Solar System and a string would be no bigger than a tree.

In the ultimate Lilliputian world of string theory, strings travel through space and interact with each other. They also vibrate, like little violin strings. To us, in the macrocosmos, a string would look like an ordinary particle, the mass, charge and other properties of which were determined by the string's vibrational state. The length of each string – about 10^{-35} metres – is assumed to be roughly the same as the shortest length that has any meaning in physics. This is known as the Planck length and is the distance at which the effects of gravity at a quantum level become significant. In fact, one of the vibrational modes of a string corresponds to the graviton, the hypothetical particle that carries the gravitational force.[2]

Scientists know that, eventually, gravity must be brought into the realm of quantum mechanics if we're to have a complete understanding of the universe at the smallest scale. String theory is one approach to achieving this – but one among many. For now, particle physicists continue their experimental and theoretical work in search of new schemes beyond the Standard Model that will lead to a deeper understanding of the world at the most fundamental level.

CHAPTER 42

A Sensitive Subject

WE CONSIDER OUR eyesight to be normal if it's 20/20, which means we can see at 20 feet what an average person can see at that distance. Some people have sharper vision. The limit to human visual acuity is generally reckoned to be about 20/10. If your vision were that good, you'd be able to make out detail at 20 feet that someone with average vision could see from no more than 10 feet away. Elsewhere in the animal kingdom, though, are eyes that are sharper still.

Birds of prey – eagles and hawks – have the keenest eyesight of all, with a visual acuity of 20/5 or 20/4. It enables them to spot small prey, such as rabbits and squirrels, at a distance of 3 kilometres or more. If our vision were that good, we'd be able see an ant crawling on the pavement from the top of a 10-storey building. Their eyesight is better than ours in other ways too. Eagles can see beyond the violet end of the spectrum, in near-ultraviolet, also known as UVA or black light. This ability makes visible to them the trails of urine left by small rodents, which may be nearby.[1]

An eagle has eyes that are almost the same size as ours even though its head is much smaller. Its eyes are fixed in their sockets and have *two* focal points – one straight ahead and the other to the side at an angle of about 45 degrees – so the bird can see forwards and sideways at the same time. The eyes are also mounted at an angle to the skull, providing much greater peripheral vision than humans have.

When it comes to seeing well at night, we tend to think of cats and owls as being the champions. Our eyes function down to about 1 lux (a measure of the amount of light per square metre). Domestic cats can see in 0.125 lux and owls in light levels ten times less than that. But in the night-vision stakes some insects have all mammals and birds beaten hands down. The eyes of dung beetles are sensitive down to one ten-thousandth of a lux, while a species of carpenter bee, *Xylocopa tranquebarica*, needs only a meagre 0.000063 lux and can fly, forage and even see in colour on moonless nights.

Most astonishing of all is the vision of that least popular of nocturnal creatures, the cockroach. Researchers at the University of Oulu in Finland set about finding just how adept these insects are at seeing in the dark – and made an astonishing discovery. Cockroaches can collect and store individual photons, or particles of light, over a period of time and build up essentially a long-exposure photograph in their heads. What would be pitch black to us, and most other creatures, can be turned into a coherent image by the remarkable roach.

For complexity, nothing surpasses the eyes of the mantis shrimp. Mounted on independently roaming stalks, each eye is equipped with three pupils (and therefore its own depth perception) and twelve to sixteen different pigments for colour vision, compared with our three (for red, green and blue). The mantis shrimp can also see in ultraviolet and is the only known animal able to detect circularly polarised light.

In Chapter 1, we talked about how low some animals can hear. But how about at the other end of the scale? The limits of human hearing are from about 20 hertz (Hz), or cycles per second, to 20,000 Hz. Sounds above and below that are said to be infrasonic and ultrasonic, respectively. In the ultrasonic range no creature can hear as high, or across such a wide range, as the greater wax moth.[2] It can detect frequencies up to at least 300,000 Hz – a fact essential to its survival. The moth is preyed upon by bats, which locate their victims using high-frequency sonar. So, the moth has evolved to be able to hear the bat's echo sounding and thus stay one step ahead in the battle for survival. There's no sound any bat can produce that the moth can't hear. What's more, its hearing is so precise that it can tell the difference between bat echolocation calls and the equally high-pitched mating calls of other greater wax moths.

As for quiet sounds – the topic of Chapter 4 – our hearing goes down to about 0 decibels (dB). We can only imagine what it would be like to have the sensitive hearing of wolves, cats or some breeds of dog. They can detect sounds as quiet as minus 15 decibels – the slightest of noises a prey animal might make many metres away. Canines and felines are among animals that can rotate their ears to better capture and locate low-volume sounds. A few bat species have such sensitive hearing they can detect the footsteps of insects walking nearby.

Our sensory world is dominated by sight and sound. If we had to rely on smell to tell us about our surroundings, we'd be lost because the olfactory sense of humans is feeble compared with that of many other species. Dogs are well known for their talent at picking up and distinguishing a wide range of odours. Their sense of smell is 10,000 to 100,000 times more acute than our own, making them invaluable in the search for missing persons or detecting illegal substances at airports. It's hardly surprising that our canine companions outclass us in the nasal department: they have up to 300 million olfactory receptors in their noses compared to our 6 million, and the parts of the brain devoted to processing odours – the olfactory bulbs – are about forty times larger in dogs than in humans.

Dogs' noses also work differently than our own. When we inhale, we smell and breathe through the same airways within our nose. When dogs inhale, a fold of tissue inside the nostril helps separate these two functions. When we exhale through our nose, the spent air goes out the same way it came in, thereby forcing out any incoming odours. When dogs exhale, the spent air leaves through slits in the sides of the nose. The way in which the exhaled air swirls out helps draw new odours into the animal's nose. It also means that dogs can sniff more or less continuously.

If length of nose is anything to go by, it's hardly surprising that elephants have the best sense of smell of any land animal. Another gauge of olfactory prowess is the number of genes devoted to scent detection. The elephant has about 2,000 olfactory receptor genes – double the number in the dog, and five times as many as we possess. An elephant can distinguish by smell alone two different ethnic groups of people in Kenya – the Maasai, who sometimes hunt elephant with spears, and the Kamba, who rarely pose a threat.

The shark's sense of smell is legendary – and, like all legends, is prone to exaggeration. A shark can't smell your blood from a mile away, nor can it detect a drop of blood in an Olympic-sized swimming pool. In any case, sharks aren't – unlike the ravening monster in *Jaws* – particularly interested in humans as prey. But it's true that some of them do have an acute olfactory sense. That's to be expected as nearly a quarter of their entire brain is devoted to processing scents.

Closely related to smell is the sense of taste. Pre-eminent among sommeliers of nature are catfish. Whereas we typically have about 5,000 taste buds on our tongues, a catfish may have 175,000 taste-sensitive cells located all over its body.

Touch is another means by which creatures can learn about their surroundings. And in this department, a strange-looking beast called the star-nosed mole excels. The star-shaped structure on their nose is covered with over 25,000 touch receptors called Eimer's organs, used to find and identify prey.

Many animals aren't restricted to the five traditional senses – sight, hearing, smell, taste and touch – with which we're familiar. Some types of snake have infrared detectors, known as pit organs, below their eyes. These house receptors that can detect heat given off by bodies up to a metre away. In the snake's brain, the heat information from the pit organs is combined with what the animal can see to help zero in on nearby prey no matter the light level.

The platypus is unique in many ways, none more so than its ability to accurately detect electrical impulses given off by the small invertebrates upon which it feeds. Around 40,000 electrosensor cells are contained in stripes in both halves of its bill. As it swims, the platypus swings its head from side to side using its electrical detectors to locate prey even in the dark, muddy water at the bottom of rivers and streams.

Other animals navigate with the help of Earth's magnetic field. Many migratory species, from salmon to sea turtles, use magnetoreception like a compass to find their way with great precision over thousands of kilometres.[3] Among birds with this ability, pigeons are especially adept. Within their beaks are structures containing magnetite (magnetic iron oxide), which give them an astonishing awareness of spatial orientation and geographical position. Bees too navigate to sources of nectar by sensing Earth's magnetism and can detect electromagnetic waves in the atmosphere that indicate an approaching thunderstorm.

CHAPTER 43

Eruption

ONE OF THE most spectacular volcanic eruptions of recent times was that of Mount St. Helens, 150 kilometres south of Seattle. At 8:32 am on Sunday, May 18, 1980, about 3 cubic kilometres of the mountain came crashing down in a massive avalanche triggered by an earthquake measuring 5.1 on the Richter scale. Nearly 600 square kilometres of forest was flattened by the blast wave or buried beneath volcanic deposits; 300 kilometres of highway were destroyed; and fifty-seven people lost their lives.

Mount St. Helens was dwarfed, however, by two other volcanic outbursts in the twentieth century. These were the eruptions of Pinatubo in the Philippines in 1981, and Novarupta in the Alaska Peninsula in 1912, which threw out an estimated 10 and 12 cubic kilometres of material, respectively. Along with the 10 billion tons of magma which spewed out of Pinatubo were 20 million tons of sulphur dioxide. This gas reacted in the atmosphere to form a global haze of sulphates and sulphuric acid, the effect of which was to reduce the amount of solar radiation reaching Earth's surface by about 10 per cent and temperatures worldwide by half a degree Celsius.

Even Pinatubo and Novarupta are small fry compared with other volcanic monsters of the past. In 1883, the island of Krakatoa in Indonesia blew apart, killing at least 40,000 people and throwing out 20 cubic kilometres of rock, ash and pumice in an explosion that was heard 3,500 kilometres away in Perth, Australia. Barograph readings showed that the pressure wave from the event circled the Earth seven times before fading out.

More violent still was the destruction of Santorini, a small island in the Aegean Sea, 3,600 years ago.[1] A central lagoon is all that remains of the great volcano that erupted in about 1620 BCE, spewing vast clouds of dust and ash, and generating a tsunami that inundated the nearby island of Crete. This event, according to some historians, triggered the demise of one of the world's great early civilisations – the Minoans. At least 30 cubic kilometres of magma, rock bombs and other debris poured out of Santorini, ranking the eruption among the top seven or eight of the past 10,000 years.

One way volcanologists classify eruptions is using the Volcanic Explosivity Index (VEI). It ranges from 0, for a non-explosive, gentle burbling of lava, to 8 in the case of the most prodigious, cataclysmic supervolcanoes.[2]

None of the volcanoes mentioned so far are supervolcanoes. Mount Etna, on the east coast of Sicily, received a rating of 3 for its 2002–3 efforts, while the 2010 air-traffic-disrupting effusions of Iceland's Eyjafjallajökull went one better with a 4. With an index of 5 comes the Mount St. Helens eruption, followed in category 6 by Pinatubo, Krakatoa and Novarupta.

Santorini, for its *circa* 1620 BCE detonation, earns an impressive 7 on the VEI scale. Only one explosion in recent times has also scored 7: that of Mount Tambora, on the island of Sumbawa in Indonesia. Tambora began to rumble in 1812, and reached a crescendo with a mind-numbing eruption in April 1815. Roughly 160 cubic kilometres of ejecta issued from Tambora's outburst, making it the largest volcanic eruption in recorded history. The vast quantities of dust and ash entering the atmosphere lowered temperatures worldwide for months, and the following year became known as the 'year without a summer'.

But neither Santorini nor Tambora were supervolcanoes. That term is reserved for eruptions involving at least *1,000* cubic kilometres of ejecta – similar in destructive capacity to a 1-kilometre-wide asteroid barrelling into the Earth. Supervolcanoes have a VEI of 8, the maximum value recognised. According to the geological record, one of these monsters explodes on average every 100,000 years. The last happened about 74,000 years ago on the island of Sumatra. Known as the Toba super-eruption, its site is marked today by Lake Toba – at 100 kilometres long by 30 kilometres wide, the largest volcanic lake in the world.

Even by supervolcano standards Toba was impressive, unrivalled by any other eruption over the past 25 million years. The amount of molten rock and other stuff it poured out came to around 2,800 cubic kilometres – more than double the volume of Mount Everest and equivalent to 10,000 volcanic eruptions of the size of Mount St. Helens going off at the same time. Toba occurred at a time when Neanderthals and more modern humans coexisted in Europe and much of Asia, and its after-effects would have brought immense hardship to our ancestors, perhaps, according to some suggestions, even pushing them to the brink of extinction.

It's generally accepted that the Toba super-eruption caused a worldwide slump in average temperatures of between 3 and 5 degrees Celsius. A blanket of ash at least 15 centimetres deep covered all of South Asia, and in places the deposition was much greater – 6 metres at one site in central India, and 9 metres in parts of Malaysia. Flora and fauna alike in Southeast Asia must have been devastated in a planet-wide die-off.

The Toba blow-out happened in the middle of a period, between 100,000 and 50,000 years ago, when the human population plummeted. This has led to the Toba catastrophe theory, according to which the effects of the eruption were so severe that the global population of *Homo sapiens* was slashed to 10,000 individuals or fewer. The die-off of vegetation, and the cooling and drying of the climate resulting from the volcanic fallout, probably altered the migratory habits of our ancestors and compelled them to adopt new methods to gain access to whatever scarce food sources were available. Ultimately, our species benefited from the ordeal of Toba; it made us tougher and more reliant on our wits and latent talents for communication and cooperation.

Earlier super-eruptions have been linked to mass extinctions, when whole swathes of life were erased within a relatively short space of time. The Permian mass extinction of 250 million years ago, which wiped out more than 90 per cent of animal species on Earth, is thought to have been tied to a colossal eruption event associated with the Siberian Traps. These Traps make up a vast region of igneous rock which formed from the outpouring of several million cubic kilometres of lava, and today covers a large part of Siberia. The activity took place over about a million years and involved individual eruptions each of which may have rivalled the Toba eruption. So dramatic were the effects on the global ecosystem that it took land life about 30 million years to recover.

The Siberian Traps are an example of what are called large igneous provinces (LIPs), vast outpourings of lava caused when giant blobs of magma in Earth's mantle burp their way to the surface. Another such catastrophic gushing of lava gave rise to the Deccan Traps, which at the time they formed, between 60 and 68 million years ago, buried most of India under molten rock. The peak of the eruption has been dated to 66 million years ago, just prior to the mass extinction at the end of the Cretaceous Period when the last of the dinosaurs and many other animals and plants disappeared from the fossil record. It's widely accepted that the dinosaurs were finished off mainly by a large asteroid collision. But environmental fallout from the Deccan Traps could have contributed to their downfall.

There's no doubt that Earth will experience more super-eruptions. Some of the supervolcanoes that have erupted in the past retain the capacity to do so again. They include one that lies at the heart of a popular and scenic tourist area.

The Yellowstone Caldera, occupying about half of Yellowstone National Park, measures some 72 kilometres by 55 kilometres and is the site of numerous past eruptions, many of them in the range of ordinary volcanoes but a few in the supervolcano class. The most recent of these gargantuan outbursts, about 640,000 years ago, is thought to have been responsible for the demise of many large mammals in North America, which choked on toxic gases or starved to death following the die-back of vegetation blanketed by a continent-wide ash cloud from the event. Other super-eruptions of the Yellowstone supervolcano happened 1.3 million and 2.1 million years ago. The average period between eruptions, geologists have determined, is about 600,000 years, which means we may be due for another one.[3]

This wouldn't be good news for America. Blanketed in a layer of ash a metre thick (and a great deal more in areas close to the eruption), the country would be rendered virtually unfit for human habitation. Midwestern states, home to much of the nation's food production and industry, would be hit especially hard. But the effects would be felt much farther afield. As in the case of the Toba eruption, global cooling, the mass dying of plants, and then the mass dying of animals and people would follow in the days, months and years after the cataclysm.

Grand Prismatic Spring in Yellowstone National Park.

Not surprisingly, folk get a little jittery when reports come through of fresh activity in the nation's favourite national park. Small earthquakes aren't unusual in Yellowstone, and the signs of geothermal activity are everywhere to be seen, from the predictable appearances of the Old Faithful geyser to the constantly bubbling, burbling, sulphurous cauldrons of hot water and mud to be found all over the Park. These crowd-pleasing features are just mild expressions of the colossal forces steadily building up below.

Between 6 and 16 kilometres beneath the picture-perfect scenery of Yellowstone is a giant magma chamber, which is slowly but surely filling with molten rock from the underlying mantle. It's an estimated 50 kilometres long, 30 kilometres wide, and 10 kilometres deep, and is fed by a magma plume that rises at a 60-degree angle from at least 660 kilometres beneath the surface. Trapped gases are steadily increasing the pressure inside the magma, and although some of that pressure is relieved on a daily basis by the various geothermal features that attract visitors to the Park, it isn't enough. At some point, the pressure inside the subterranean chamber will reach a critical level, the overlying rock will be split apart, and the gas-laden magma will erupt explosively over a wide area at the surface. On that fateful day, more than 1,000 cubic kilometres of magma might burst forth, bringing chaos to North America and the wider world.

The good news is that the Yellowstone Caldera is one of the most closely monitored volcanically active regions on Earth. Scientists with the USGS, University of Utah and National Park Service said recently that they 'see no evidence that another ... cataclysmic eruption will occur at Yellowstone in the foreseeable future'.

CHAPTER 44

The Methuselah Syndrome

MANY CLAIMS HAVE been made for extreme human longevity. But the oldest known person based on official records was Frenchwoman Jeanne Calment, who died in 1997 at the age of 122. Few mammals live longer on average than we do. One of the exceptions is the bowhead whale, a denizen of Arctic seas.

In 2007, a 50-ton bowhead was killed off the Alaskan coast by Inupiat whalers using a bomb lance – a spear-like projectile carrying an explosive which detonates once it's embedded inside its prey. When the whale was cut open it was found to contain the head of a much older bomb lance manufactured between 1879 and 1885.[1] This led to an estimate of the animal's age of between 120 and 130 years. Spurred by the discovery, scientists began measuring the ages of other bowheads and found one to be as old as 211.

This suggests that some larger whales have evolved special mechanisms that protect them against ageing and life-limiting diseases such as cancer. To uncover the secrets of their longevity, researchers from the US and UK in 2015 mapped the bowhead's genome – the entire set of DNA instructions found within each cell. It was the first time that the genome of a large cetacean had been sequenced. The whale's genetic sequence was compared with that of nine other mammals, including cows, rats and humans.

The comparative analysis revealed mutations in two genes that help explain the bowhead's ability to live longer. One of these mutations enables the whale to be better at repairing damage to its DNA. The other is linked both to DNA repair and conferring greater resistance to cancer. The bowhead also has a low metabolic rate related to the cold environment in which it lives. Based on all these factors, scientists estimate the whale's maximum lifespan is around 268 years.

Longer lived still is another dweller of frigid Arctic waters – the Greenland shark. Its great longevity was suspected as long ago as the 1930s. At that time a fisheries biologist in Greenland tagged several hundred of these large, slow-swimming, deep-water fish and found that they grow extremely slowly – as little as 1 centimetre a year. It was a sure sign that their lives can stretch across centuries, given that they sometimes reach more than 7 metres in length.

More recently, marine biologist John Steffensen at the University of Copenhagen set out to pinpoint more accurately just how old Greenland sharks can get. He examined a piece of backbone from a specimen caught in the North Atlantic, hoping to count growth rings that would give the age of the animal, but none were to be seen. He then turned to Jan Heinemeier, a specialist in radiocarbon dating at Aarhus University. Heinemeier suggested that the lenses of the sharks' eyes might hold a clue to how long the animals had lived.

For the next few years Steffensen and a graduate student, Julius Nielsen, collected lenses from dead Greenland sharks, most of which had become trapped in trawling nets set for other types of fish. Then they looked in the lenses for concentrations of carbon-14. This radioactive form of carbon was one of the by-products of hydrogen bomb testing in the 1950s, which, by the early 1960s, had found its way into ocean ecosystems. Inert body parts, including eye lenses, of animals born during this time have elevated amounts of the isotope. Thanks to this dating method, the two Danish researchers were able to show that two of the sharks in their study were born after the 1960s, while another was born in about 1963.

From these accurately determined ages, and knowing that newborn Greenland sharks are about 42 centimetres long, they were able to plot a growth curve showing how a shark's age correlated with its length. The oldest of the twenty-eight specimens they examined was a 5-metre-long female. Her age was astonishing: 392 plus or minus 120 years. Even at the lower end of that range she was easily the oldest vertebrate on record. What's more, the researchers found, given that most pregnant female Greenland sharks are close to 4 metres in length, they don't even have young until they're at least 150 years old.[2]

There are older creatures in the sea – if we allow invertebrates (animals without backbones) into the reckoning. Meet Ming the quahog or chowder clam. No radioactive dating is needed to estimate the age of a bivalve mollusc such as this – you simply count the annual growth rings inside its shell. In 2006, a specimen of quahog was dredged off the coast of Iceland that had an astonishing number of growth rings. To begin with scientists counted 405 of them, earning the clam its nickname after the Chinese dynasty in which it was born. A later determination corrected this figure to 507, fixing Ming's birth in the year 1499 – four years before Leonardo da Vinci began work on the *Mona Lisa*. In one of the ironies of scientific research, Ming's shell was pried open, killing the oldest known animal on the planet, in order to accurately determine its age.[3]

On land, the most elderly known animal still alive is Jonathan the Giant Tortoise. Born in the Seychelles, he was fully mature – meaning that he was at least fifty years old – when brought to his present home, St Helena, in 1882. Assuming a birth year of 1832, he is now 191. But Jonathan has some way to go before overtaking the oldest tortoise whose age was fairly well established. Adwaitya ('one and only' in Sanskrit) was one of four tortoises given to Clive of India in 1757 following his victory at the Battle of Plassey. About twenty years later he was transferred to the Alipore Zoological Gardens in Kolkata, where he lived until his death in 2006 aged about 255.

To venture into the extremes of individual longevity we have to turn to the world of plants. Some types of tree can live for not just centuries but millennia. There is a sacred fig (*Ficus religiosa*) in the grounds of the Mahamewna Gardens in Anuradhapura, Sri Lanka, which was planted in 288 BCE. At 2,310 years of age it is the oldest known living human-planted tree in the world. The President, in Sequoia National Park, California, is approximately 3,200 years of age. The ring count of Methuselah, a bristlecone pine, in the White Mountains of California, reveal it to be 4,854 years old, so that it had already been alive for three centuries when the Great Pyramid of Giza was built.[4]

If five millennia is about the limit, as far as we know, for *individual* living things, it's far from being so for other forms of organism on the planet. Some types of plant live in what are called clonal colonies in which every member is genetically identical to every other. If such colonies are considered to be a single living entity, then their lifespans can be enormous. A colony of seagrass in the Mediterranean Sea off the coast of Ibiza is believed to be at least 12,000 years old, while the age of a colony of shrub known commonly as King's lomatia, in Tasmania, has an estimated age of at least 43,600 years.

Descending to the realm of microbes, longevity can stretch even further. Some endoliths – microorganisms that inhabit tiny pores inside rocks and minerals, living purely on chemicals in their surroundings – have been alive longer than *Homo sapiens*. In 2013, researchers found evidence of endoliths beneath the ocean floor that may be millions of years old, with a generation time of 10,000 years. Yet even this is surpassed by creatures that can survive for incredible periods of time in a state of complete dormancy, or suspended animation, before resuming their metabolic activity. Scientists have reported the revival of spores, trapped in amber (fossilised resin), after 40 million years, and from salt deposits in New Mexico after 250 million years. The latter have survived since the end of the Permian period, before the age of dinosaurs, and are the oldest known living things on Earth.

Finally, there may be immortal beings – creatures that never age or, in some cases, are able to rejuvenate. Some species of *Hydra*, small freshwater organisms with a tubular body, tentacles that carry stinging cells, and an adhesive foot, appear never to die of old age. If injured, or even cut in half, they will regenerate. In 1998, Daniel Martinez at the University of Arizona became the first to claim that *Hydra* are biologically immortal and therefore prove the existence of 'non-senescing' life-forms. Although his results attracted controversy, research carried out since has supported the notion that some organisms can, at least in theory, live forever.

CHAPTER 45

Great Survivors

THE FIRST APOLLO astronauts to return from the Moon came back to cheering crowds – and three weeks of isolation in case they'd unwittingly brought any deadly pathogens back from the lunar surface. No one seriously expected there'd be life on the Moon but the risk of unleashing an alien bug, to which we might have zero resistance, was too big to take. Who knew under what extreme conditions life elsewhere in the universe might have managed to adapt?

Since the time of Apollo, scientists have learned a great deal more about so-called 'extremophiles' – creatures that, to us, inhabit the most incredibly hostile places imaginable. Life on our own world, it turns out, has radiated into every conceivable – and in some cases almost inconceivable – ecological niche.

Some hardy animals are familiar. Camels can go a week or more without water in temperatures as high as 49 °C. Male emperor penguins in Antarctica stand, twenty-four hours a day, for two months in temperatures that routinely drop to −40 °C while incubating an egg that's balanced on their feet. A colony survives under these conditions by huddling together in huge groups and rotating individuals from the exposed outer margins to the middle so that every member has a chance to 'warm' up. Wood frogs take a different approach to dealing with the cold: their bodies freeze and remain in suspended animation until the spring thaw. They survive being frozen by accumulating glucose, a cryoprotectant, in their tissues.

Over the past few decades, scientists have uncovered a vast array of extremophiles living in what previously had been thought of as uninhabitable places. Thermophiles like it hot, and hyperthermophiles grow and multiply happily in temperatures that would cook most living things within a few seconds. In 1997, researchers investigating a 'black smoker' hydrothermal vent on the Atlantic Ocean floor discovered a type of microbe they called *Pyrolobus fumarii* (literally the 'firelobe of the chimney') surviving at temperatures of up to 113 °C.

More recently, that record has been broken by *Methanopyrus kandleri*, which, like *Pyrolobus*, is a type of archaea – a single-celled organism similar to a bacterium but that's evolved down a completely separate evolutionary pathway. Many archaea have adapted to life in extreme environments. Their cell walls differ in structure from those of bacteria, making them more stable in hostile conditions, and their internal chemistry lets them function normally in places that would instantly kill other life-forms. *Methanopyrus kandleri* can survive temperatures of up to 122 °C, multiplies best at around 98 °C, and 'freezes' or solidifies and stops growing below about 90 °C.[1]

At the other extreme, psychrophiles flourish in bone-chilling locales. These are creatures to which polar ice, glaciers, snowfields and permafrost are home. Among cold-loving insects are the nocturnal ice crawlers, with short bodies, cockroach-like heads, and long antennae, which thrive in just above freezing temperatures and die at 10 °C. The wingless midge, *Belgica antarctica*, can tolerate not only being frozen but also exposure to salt and strong ultraviolet at the same time. It's a polyextremophile – a life-form that thrives in multiple different extreme environments – and the fact that it has the smallest known genome of any insect is thought to be an adaptation of this.

Also among the ranks of psychrophiles are certain kinds of algae, bacteria, lichen and fungi. Some lichen have been recorded photosynthesising at temperatures as low as −24 °C and microbial activity has been measured in soils below −39 °C.

It seems impossible that anything could live inside rock deep underground, where there's no air, light or water. Yet endoliths manage to eke out a living within the tiny pores and fissures that exist between mineral grains, sometimes a mile or more underground. In the case of autoendoliths they require nothing other than a small chemical diet to sustain their growth and metabolism. Some may undergo cell division only every century or so, and in 2013 scientists reported evidence of endoliths beneath the ocean floor that may be millions of years old and that reproduce once every 10,000 years.

Extremophiles have been found in the highly acidic water bubbling out of the ground in some areas of geothermal activity. Thermoacidophiles abound in the nightmare environment of near-boiling hot springs in the Norris Geyser Basin, Mud Volcano and Sulphur Caldron areas of Yellowstone National Park. The steaming, sulphurous brew of these ponds may be as hot as 90 °C and have a pH (a measure of acidity and alkalinity) similar to that of battery acid. But within these hellish waters dwell microbes such as those of the genus *Sulfolobus* and the species appropriately named *Acidianus infernus*. Researchers are interested in these incredibly tough organisms not just as curiosities in themselves but as a source of thermostable enzymes (biological catalysts) for use in the food, textile and paper industries, as well as scientific research and diagnostics.

Among the hardiest living things on the planet are tardigrades, or water bears – strange-looking, eight-legged micro-animals, which, with a maximum length of half a millimetre, are barely visible to the naked eye.[2] They've been found almost literally everywhere on Earth from deserts to glaciers, and from hot springs to the frozen high peaks of the Himalayas. There are few environments in which they can't live, and when conditions get especially harsh they have the ability to survive by entering a dried-up, deathlike state known as cryptobiosis. In this condition they can remain for decades or more until exposure to water causes them to become active again.

A tardigrade, or 'water bear' – one of the
hardiest organisms on Earth.

There may even be tardigrades on the Moon – not indigenous ones but visitors from Earth accidentally released following the crash landing of an Israeli probe. *Beresheet* was to have been the first privately funded lunar lander, its goal: to ensure there was a back-up in case all life on Earth was destroyed. Among its payload was a DVD-sized archive containing 30 million pages of information (including all of the English Wikipedia), human DNA samples – and thousands of dehydrated tardigrades. The little creatures may well have survived *Beresheet*'s failed attempt at a soft touchdown, but the absence of lunar water means there's no chance of them reanimating and populating the Moon with its first colony of terrestrial émigrés.

Some other worlds in space, though, look much more promising from a biological standpoint, and the discovery of a wide diversity of extremophiles on Earth has encouraged the belief that life may be fairly common throughout the universe. Mars was once a warmer, wetter place than it is today so that life may have developed there in the past. If so, it may have adapted, over billions of years, as conditions worsened and – just possibly – still exist there today in more sheltered spots, such as deep underground.

To better understand the chances of finding life elsewhere in the Solar System, and beyond, scientists seek out terrestrial analogues – locations on Earth that have features in common with alien environments. One of these is Pitch Lake on the island of Trinidad, which a friend of mine, the astrobiologist Dirk Schulze-Makuch, has spent time exploring. Pitch Lake is exactly that – a 109-acre expanse of gooey, black, natural asphalt whose contents have been excavated to pave roads and airport runways around the world, including the roadway in front of Buckingham Palace in London and the runway at La Guardia Airport in New York. A less promising place to search for life of any kind you could hardly imagine. Yet within the sticky tar of Pitch Lake are tiny pockets of water, within which have been found colonies of microbes.[3] In fact, these miniature watery bubbles contained within the asphalt are the smallest known isolated ecosystems on Earth.

Pitch Lake may be the closest thing we have on this planet to the surface of Saturn's giant moon Titan. On Titan there are lakes, rivers and even small seas full of hydrocarbons, of which asphalt is a variety. Water, from deep underground, may occasionally be brought to the surface, to mix with the hydrocarbons there, by asteroid collisions. So, it's conceivable that the conditions for life may exist on the Saturnian moon, unlikely as it may appear, just as they do in the tarry expanse at La Brea.

Some extremophiles can survive even in the airless, radiation-blasted vacuum of space. In 2015, a robotic arm mounted a box of microbes outside Japan's Kibo lab, which forms part of the International Space Station. The bacteria in the box were totally at the mercy of high-energy ultraviolet, X-rays and gamma rays, which constantly strafe the region above our atmosphere. After three years, the experiment was recovered and brought back to Earth. In 2020, Japanese scientists published their findings.

One of the species of germ involved was *Deinococcus radiodurans*, which was already known to be radiation-resistant because its genes code for special proteins that can repair DNA. The cells on the outer layers in the experiment had been killed but these dead cells, it turned out, had shielded their comrades further inside from irreparable DNA damage. The combined effect of the protective sacrificial layer and the inherent genetic defence mechanism of *D. radiodurans* kept the bulk of the microbes alive for three years, despite the intense blitz going on all around them.

The ability of this organism to survive in space suggests that some living things, at least, might be able to make journeys lasting several years or more between worlds aboard debris thrown from the surface of one of the worlds by an asteroid collision. The question has been asked: could microscopic life have been transported, billions of years ago, from Earth to Mars, thereby seeding the fourth planet with some of our primitive ancestors? Or, did perhaps life emerge first on Mars, when the Solar System was young, and then find its way here so that, in some sense, we're the descendants of Martians?

Conclusion

WE'RE ATTRACTED TO the strange, the exotic, the extreme. We're intrigued by adventures beyond the normal, by explorations beyond the limit of what was previously attainable, because they catapult us out of our ordinary lives. New records in human endeavour or nature will always make headlines. It's part of the human condition to seek new ground, to break through old barriers.

But the urge to go one better than what came before is motivated by more than just a desire to see records broken for the sake of it. Very often there are practical reasons to want barriers transcended. We seek new ways to travel faster or more economically, to produce more energy with less harm to the environment, to discover lighter, stronger, better insulating or conducting materials. In physics and astronomy, we want to know what came first, what are the ultimate constituents of matter and energy, and how nature behaves under the most extraordinary conditions. For as long as our species survives we'll continue to explore the limits of the possible – the science of extremes.

Acknowledgements

I'm tremendously grateful, as always, for the loving support and encouragement of my family, especially my wife, Jill.

Thank you, too, to the wonderful folk at Oneworld who have made this book possible – first and foremost my editor, Sam Carter, for his guidance and insights.

References and Related Videos

1. HOW LOW CAN YOU GO?

1 'Greatest vocal range, male', Guinness World Records website.
http://www.guinnessworldrecords.com/world-records/3000/greatest-vocal-range-male

2 'The Montreal Symphony Orchestra's new octabass has arrived', Robert Rowat, CBC, June 16, 2016.
https://www.cbcmusic.ca/posts/11590/montreal-symphony-orchestra-new-octobass

3 'Infrasound linked to spooky effects', NBC News, September 2003.
https://www.nbcnews.com/id/wbna3077192

4 'Disappearing Homing Pigeon Mystery Solved', Kathryn Knight, *Journal of Experimental Biology* 216 (4), February 15, 2013.

5 'Can Animals Predict Disaster?', PBS, June 5, 2008.
https://www.pbs.org/wnet/nature/can-animals-predict-disaster-introduction-2/134/

6 'Interpreting the "Song" of a Distant Black hole', NASA.
https://www.nasa.gov/centers/goddard/universe/black_hole_sound.html

2. SL-0-0-0W!

1 'A New Way to Laser-Cool Molecules', Nicholas R. Hutzler, *Physics* 13 (89), June 3, 2020.
https://physics.aps.org/articles/v13/89

2 'Observing the Rarest Decay Process in Nature', Purdue University.
https://science.purdue.edu/xenon1t/?p=1287

3 'The universe's biggest gear reduction! GOOGOL to 1', Daniel de Bruin.
https://www.youtube.com/watch?v=nFsIB0AcVmM

3. BRILLIANT

1   'Scientists face down "Godzilla", the most luminous star known', *Nature* 610 (7930): 10, October 6, 2022.

2   'Found: The Most Powerful Supernova Ever Seen', Lee Billings, *Scientific American*, January 14, 2016.

https://www.scientificamerican.com/article/found-the-most-powerful-supernova-ever-seen/

3   '1 billion suns: World's brightest laser sparks new behavior in light', Scott Schrage, University of Nebraska.

https://news.unl.edu/newsrooms/today/article/1-billion-suns-world-s-brightest-laser-sparks-new-behavior-in-light/

4. SHHH

1   *4'33"*, John Cage, Performance at the Barbican by the BBC Symphony Orchestra.

https://www.youtube.com/watch?v=yoAbXwr3qkg

2   'Experience the Quietest Place on Earth', Margaret Cirino, Regina G. Barber and Gabriel Spitzer, NPR, August 26, 2022.

https://www.npr.org/2022/08/25/1119484767/experience-the-quietest-place-on-earth

3   'Confirmed: In Space No One Can Hear You Scream', Tim Pilgrim, Brunel University.

https://www.brunel.ac.uk/news-and-events/news/articles/Confirmed-In-space-no-one-can-hear-you-scream

4   'NASA's Perseverance Rover Captures the Sounds of Mars', NASA Jet Propulsion Laboratory.

https://www.youtube.com/watch?v=GHenFGnixzU

5   'Where Sound Goes to Die', Microsoft.

https://news.microsoft.com/stories/building87/audio-lab.php

5. UP TO ELEVEN

1   'The Audiological Health of Horn Players', Wayne J. Wilson, Ian O'Brien and Andrew P. Bradley, *Journal of Occupational and Environmental Hygiene*, 10:11 (590–596), 2013.

2   'Large European Acoustic Facility', European Space Agency.

https://www.esa.int/Enabling_Support/Space_Engineering_Technology/Test_centre/Large_European_Acoustic_Facility_LEAF

3 'ZWRRWWWBRZR: That's the sound of the prop-driven XF-84H, and it brought grown men to their knees', Stephan Wilkinson, *Air & Space*, Smithsonian Institution.

http://www.airspacemag.com/how-things-work/zwrrwwwbrzr-4846149

4 'How Krakatoa made the biggest bang', *The Independent*, May 3, 2006.

https://www.independent.co.uk/news/science/how-krakatoa-made-the-biggest-bang-5336165.html

5 'What's the loudest a sound can be?', *BBC Science Focus*, December 29, 2020.

https://www.sciencefocus.com/science/whats-the-loudest-a-sound-can-be/

6. ALMOST 0 K

1 'Record low surface air temperature at Vostok Station, Antarctica', British Antarctic Survey, December 27, 2009.

https://www.bas.ac.uk/data/our-data/publication/record-low-surface-air-temperature-at-vostok-station-antarctica/

2 'Shadowy moon crater coldest spot yet measured', CBC News, September 18, 2009.

https://www.cbc.ca/news/science/shadowy-moon-crater-coldest-spot-yet-measured-1.851001

3 'The Boomerang Nebula: The Coldest Region of the Universe?', Raghvendra Sahai and Lars-Åke Nyman, Harvard University, October 1997.

https://ui.adsabs.harvard.edu/abs/1997ApJ...487L.155S/abstract

4 'CUORE has the coldest heart in the known universe', *CERN Courier*, November 27, 2014.

https://cerncourier.com/a/cuore-has-the-coldest-heart-in-the-known-universe/

5 'NASA's Cold Atom Lab Takes One Giant Leap for Quantum Science', NASA, June 13, 2020.

https://www.nasa.gov/feature/jpl/nasas-cold-atom-lab-takes-one-giant-leap-for-quantum-science

6 'New record set for lowest temperature – 38 picokelvins', Phys.org, October 13, 2021.

https://phys.org/news/2021-10-coldest-temperature38-picokelvins.html

7. HOT TOPIC

1 'Planet WASP-12b is on a death spiral, say scientists', Princeton University.

https://www.princeton.edu/news/2020/01/07/planet-wasp-12b-death-spiral-say-scientists

2 'Astronomers Discover a Giant Planet Hotter Than Most Stars', *Scientific American*, June 5, 2017.

https://www.scientificamerican.com/article/feeling-hot-hot-hot-astronomers-discover-a-giant-planet-hotter-than-most-stars/

3 'The Most Extreme Stars in the Universe', Jake Parks, *Astronomy*, September 23, 2020.

https://astronomy.com/magazine/news/2020/09/the-most-extreme-stars-in-the-universe

4 'Another world record for China's EAST tokamak', *Nuclear Engineering International*, April 18, 2023.

https://www.neimagazine.com/news/newsanother-world-record-for-chinas-east-tokamak-10768385

5 'Lawrence Livermore National Laboratory achieves fusion ignition', Lawrence Livermore National Laboratory, December 14, 2022.

https://www.llnl.gov/news/lawrence-livermore-national-laboratory-achieves-fusion-ignition

6 'European researchers achieve fusion energy record', *EUROfusion News*, February 9, 2022.

https://euro-fusion.org/eurofusion-news/european-researchers-achieve-fusion-energy-record/

7 'Two CU Physics Professors Part of Team That Created World's Hottest Temperature Matter in Atom Smasher', *CU Boulder Today*, University of Colorado, February 16, 2010.

https://www.colorado.edu/today/2010/02/16/two-cu-physics-professors-part-team-created-worlds-hottest-temperature-matter-atom

8 'CERN physicists break record for hottest manmade material', Phys. org, August 16, 2012.

https://phys.org/news/2012-08-cern-physicists-hottest-manmade-material.html

8. SPHERE

1 'Roundest objects in the world created', Devin Powell, *New Scientist*, July 1, 2008.

https://www.newscientist.com/article/dn14229-roundest-objects-in-the-world-created/

2 'A Pocket of Near-Perfection', *NASA Science*, April 26, 2004.
https://science.nasa.gov/science-news/science-at-nasa/2004/26apr_gpbtech

3 'The Sun's almost perfectly round shape baffles scientists', *Astronomy*, August 17, 2012.
https://www.astronomy.com/news/2012/08/the-suns-almost-perfectly-round-shape-baffles-scientists

4 'Kepler 11145123 Is Most Spherical Natural Object Ever Seen, Astronomers Say', *Sci News*, November 18, 2016.
https://www.sci.news/astronomy/kepler-11145123-most-spherical-natural-object-04378.html

5 'Supernova Leaves Behind Mysterious Object', *Space Daily*, July 14, 2006.
https://www.spacedaily.com/reports/Supernova_Leaves_Behind_Mysterious_Object_999.html

9. BEYOND THE SUPERBALL

1 'Why Physicists Love Super Balls', Joel Shurkin, *Inside Science*, May 22, 2015.
https://www.insidescience.org/news/why-physicists-love-super-balls

2 'Record-breaking Steel Could Be Used for Body Armor, Shields for Satellites', University of California, San Diego, April 5, 2016.
https://jacobsschool.ucsd.edu/news/release/1915

10. TESLA MAX

1 'How strong are neodymeium magnets?', Mario Gudec, YouTube video.
https://www.youtube.com/watch?v=TUuI58qwEvI

2 'Floating Frogs', *Science*, April 14, 1997.
https://www.science.org/content/article/floating-frogs

3 'Fermilab achieves 14.5-tesla field for accelerator magnet, setting new world record', Fermilab, July 13, 2020.
https://news.fnal.gov/2020/07/fermilab-achieves-14-5-tesla-field-for-accelerator-magnet-setting-new-world-record/

4 'With mini magnet, National MagLab creates world-record magnetic field', National High Magnetic Field Laboratory, June 12, 2019.
https://nationalmaglab.org/news-events/news/lbc-project-world-record-magnetic-field/

5   'China Sets World Record in Steady High Magnetic Field Research', Chinese Academy of Sciences, August 15, 2022.

https://english.cas.cn/newsroom/cas_media/202208/t20220815_311479. shtml

11. STOP RIGHT THERE

1   'Space Shuttle Thermal Tile Demonstration', Roscket Tasartir, YouTube video.

https://www.youtube.com/watch?v=Pp9Yax8UNoM

12. THE PERSISTENCE OF SOUND

1   'Music project launches at iconic Cupar silo', Michael Alexander, *The Courier*, May 20, 2016.

https://www.thecourier.co.uk/fp/entertainment/music/173694/music-project-launches-iconic-cupar-silo/

2   'Testing the World's Longest Echo', Tom Scott, Facebook video.

https://www.facebook.com/watch/?v=1418903461877118

3   'The Acoustics of the Auditorium of the Royal Albert Hall Before and After Redevelopment', R. A. Metkemeijer, Adviesbureau Peutz & Associés B.V.

https://peutz.nl/sites/peutz.nl/files/publicaties/Peutz_Publicatie_RM_ IOA_05-2002.pdf

4   'Whispering Galleries', Acoustical Surfaces, Inc., February 24, 2020.

https://www.acousticalsurfaces.com/blog/acoustics-education/ whispering-galleries

13. GEE WHIZ

1   'Physics of the Flip Flap Rollercoaster', Math! Science! History!, January 19, 2020.

https://mathsciencehistory.com/2020/01/19/physics-of-the-flip-flap-rollercoaster/

2   'High G Research on the Johnsville Centrifuge', Southeastern Pennsylvania Cold War Historical Society, YouTube video.

https://www.youtube.com/watch?v=bTFu1v_sxKl

3   'The Rocket Sled Trials of Colonel John Stapp', The History Guy, YouTube video.

https://www.youtube.com/watch?v=JHGJ_y4aJll

14. MEGA-WORLDS

1 'HR 8799 Super-Jupiters' Days Measured for the First Time', W. M. Keck Observatory, July 29, 2021.

https://www.keckobservatory.org/kpic/

2 'Smallest-ever Star Seen by Scientists', University of Cambridge, July 12, 2017.

https://www.cam.ac.uk/research/news/smallest-ever-star-discovered-by-astronomers

3 'First ever image of a multi-planet system around a Sun-like star', European Southern Observatory, July 22, 2020.

https://www.eso.org/public/images/eso2011b/

15. STELLAR SUPERSTARS

1 'The Largest Star (Stephenson 2-18)', SEA, February 11, 2021.

https://www.youtube.com/watch?v=W9ME-WBIkeM

2 'Sharpest Image Ever of Universe's Most Massive Known Star', NOIRLab, National Science Foundation, August 18, 2022.

https://noirlab.edu/public/news/noirlab2220/

16. BIG NEWS

1 'Laniakea: Our Home Supercluster', *Sky & Telescope*, September 3, 2014.

https://skyandtelescope.org/astronomy-news/laniakea-home-supercluster-09032014/

2 'New galaxy supercluster spotted', *Nature India*, July 14, 2017.

https://www.nature.com/articles/nindia.2017.83

3 'What Is the Biggest Thing in the Universe?', Science ABC, July 8, 2022.

https://www.scienceabc.com/nature/universe/what-is-the-biggest-thing-in-the-universe.html

17. FAR, FAR AWAY

1 'The Most Distant Milky Way Stars', *Sky & Telescope*, July 9, 2014.

https://skyandtelescope.org/astronomy-news/the-most-distant-milky-way-stars-070920142/

2 'A Planetary Detection in Andromeda?', Paul Gilster, Centauri Dreams, June 11, 2009.

https://www.centauri-dreams.org/2009/06/11/a-planetary-detection-in-andromeda/

3 'Meet Earendel: Hubble telescope's most distant star discovery gets a Tolkien-inspired name', Space.com, April 1, 2022.

https://www.space.com/hubble-most-distant-star-tolkien-name-earendil

4 'JWST's Newfound Galaxies Are the Oldest Ever Seen', *Scientific American*, April 13, 2023.

https://www.scientificamerican.com/article/jwsts-newfound-galaxies-are-the-oldest-ever-seen/

18. KA-BOOM!

1 'Russia releases secret footage of 1961 Tsar Bomba hydrogen blast', Reuters, August 28, 2020, YouTube video.

https://www.youtube.com/watch?v=YtCTzbh4mNQ

2 'Ophiuchus Galaxy Cluster', NASA, February 27, 2020.

https://www.nasa.gov/mission_pages/chandra/images/ophiuchus-galaxy-cluster.html

19. RACING THROUGH THE COSMOS

1 'NASA Probe, Fastest Object Built by Humans, Passes Sun at Record-Breaking 364,621 mph', *Newsweek*, November 22, 2021.

https://www.newsweek.com/nasa-parker-solar-probe-fastest-object-built-humans-passes-sun-record-breaking-364621-mph-1651815

2 'We Just Found the Fastest Star in the Milky Way, Travelling at 8% the Speed of Light', ScienceAlert, August 13, 2020.

https://www.sciencealert.com/the-fastest-star-in-the-galaxy-zooms-as-high-as-8-percent-of-the-speed-of-light

20. HOW DENSE CAN YOU GET?

1 'Osmium weighs in', Gregory Girolami, *Nature Chemistry*, October 23, 2012.

https://www.nature.com/articles/nchem.1479

2 'Unobtanium, Neutronium and Metallic Hydrogen', University of Warwick, October 30, 2020.

https://warwick.ac.uk/fac/sci/physics/research/astro/people/stanway/sciencefiction/cosmicstories/unobtanium_neutronium_and/

3 'A quark star may have just been discovered', *Advanced Science News*, November 4, 2022.

https://www.advancedsciencenews.com/a-quark-star-may-have-just-been-discovered/

21. BLACK

1 'Fifty shades of black', *Physics World*, November 5, 2015.
   https://physicsworld.com/a/fifty-shades-of-black/

2 'About Vantablack', Surrey NanoSystems.
   https://www.surreynanosystems.com/about/vantablack

3 'The "blackest" black: How a color controversy sparked a years-long art feud', CNN, August 20, 2021.
   https://edition.cnn.com/style/article/blackest-black-ink-culture-hustle/index.html

4 'Black Beast: Vantablack light-absorbing paint meets BMW', BMW.
   https://www.bmw.com/en/design/the-bmw-X6-vantablack-car.html

5 'MIT engineers develop "blackest black" material to date', *MIT News*, September 12, 2019.
   https://news.mit.edu/2019/blackest-black-material-cnt-0913

22. REFLECT ON THIS

1 'The Leviathan's Legacy: the story of the Birr Castle telescope', *BBC Sky at Night* magazine, March 16, 2017.
   https://www.skyatnightmagazine.com/space-science/the-leviathans-legacy/

2 'NASA's James Webb Space Telescope: Optics', Space Telescope Science Institute.
   https://www.stsci.edu/files/live/sites/www/files/home/jwst/about/history/flyers/_documents/JWST-Optics.pdf

23. SLIP SLIDING

1 'Molecular Insight into the Slipperiness of Ice', Mischa Bonn, Daniel Bonn, et al., *Journal of Physical Chemistry Letters* 2018, 9, 11, 2838–2842, May 9, 2018.
   https://pubs.acs.org/doi/full/10.1021/acs.jpclett.8b01188#

2 'Hagfishes: how much slime can a slime eel make?', Emily Osterloff, Natural History Museum.
   https://www.nhm.ac.uk/discover/how-much-slime-can-a-hagfish-make.html

3 'Pitcher Plant Inspires Super Slippery Surface', *Chemical & Engineering News*, September 21, 2011.
   https://cen.acs.org/articles/89/web/2011/09/Pitcher-Plant-Inspires-Super-Slippery.html

24. SLO-MO FLOW

1 'The Great Boston Molasses Flood: why the strange disaster matters today', Sarah Betencourt, *The Guardian*, January 13, 2019.

https://www.theguardian.com/us-news/2019/jan/13/the-great-boston-molasses-flood-why-it-matters-modern-regulation

2 'How does glass change over time?', Lori Baker, MIT School of Engineering, December 14, 2010.

https://engineering.mit.edu/engage/ask-an-engineer/how-does-glass-change-over-time

25. POISON MOST DEADLY

1 'Agatha Christie to the Rescue', Karen de Witt, *Washington Post*, June 24, 1977.

https://www.washingtonpost.com/archive/lifestyle/1977/06/24/agatha-christie-to-the-rescue/d1c53130-0885-4f2e-a1f1-0143db81244e/

2 'The pitohui bird contains deadly batrachotoxin', Joe Schwarcz, McGill University, April 26, 2018.

https://www.mcgill.ca/oss/article/did-you-know/pitohui-bird-contains-deadly-batrachotoxin

26. YOU'RE SO SWEET

1 'The Pursuit of Sweet', Jessie Hicks, Science History Institute, May 2, 2010.

https://www.sciencehistory.org/distillations/the-pursuit-of-sweet

27. STICKY PROBLEMS

1 'How Neanderthals made the very first glue', University of Leiden, August 11, 2017.

https://www.universiteitleiden.nl/en/news/2017/08/first-glue-neanderthals

2 'How do gecko lizards unstick themselves as they move across a surface?', Kellar Autumn, *Scientific American*, September 29, 2003.

https://www.scientificamerican.com/article/how-do-gecko-lizards-unst/

3 'Harry Coover, Super Glue's Inventor, Dies at 94', Elizabeth A. Harris, *The New York Times*, March 7, 2011.

https://www.nytimes.com/2011/03/28/business/28coover.html

28. PHEW

1 'The World's Favorite Scent Is Vanilla, According to Science', Elizabeth Gamillo, *Smithsonian*, April 6, 2022.

https://www.smithsonianmag.com/smart-news/vanilla-is-earths-most-preferred-smell-regardless-of-cultural-background-180979870/

2 '*Titan arum*', Royal Botanic Gardens Kew.

https://www.kew.org/plants/titan-arum

3 'Smelliest cheese honour', Patrick Barkham, *The Guardian*, November 26, 2004.

https://www.theguardian.com/uk/2004/nov/26/research.highereducation

4 'Things I Won't Work With: Thioacetone', *Science*, June 11, 2009.

https://www.science.org/content/blog-post/things-i-won-t-work-thioacetone

29. NEXT TO NOTHING

1 'Watch a "ballooning" spider take flight', *Science Magazine*, April 2, 2018, YouTube video.

https://www.youtube.com/watch?v=JrS0igctMi0

2 'May 1931: Publication of the Creation of the First Aerogel', American Physical Society, May 2021, 30 (5).

https://www.aps.org/publications/apsnews/202105/history.cfm

30. BUBBLES: BIG, BEAUTIFUL AND BIZARRE

1 'June 28, 1984 Bubble Master Eiffel Plasterer on David Letterman', YouTube video.

https://www.youtube.com/watch?v=OynShAgexOM

2 'Chemical analysis of gaseous bubble inclusions in amber: The composition of ancient air?', Robert A. Berner and Gary P. Landis, *American Journal of Science* 287 (757–762), October 1987.

https://www.ldeo.columbia.edu/~dmcgee/Carbon/Readings_files/Berner_Landis_87.pdf

31. ACID TEST

1 'Why don't our digestive acids corrode our stomach linings?', William K. Purves, *Scientific American*, October 20, 2003.

https://www.scientificamerican.com/article/why-dont-our-digestive-ac/

2   'Fluorosulphuric acid', American Chemical Society, May 2, 2016.
     https://www.acs.org/molecule-of-the-week/archive/f/fluorosulfuric-acid.
     html

3   'The Strongest Acid in the World: Fluoroantimonic acid', Chemicalforce,
     YouTube video.
     https://www.youtube.com/watch?v=UWBNcMyfiGQ

32. CLEARLY THE BEST

1   'What determines whether a substance is transparent?', S. M. Thomas,
     Scientific American, October 21, 1999.
     https://www.scientificamerican.com/article/what-determines-whether-a/

33. RARE

1   'Scientists uncover the fundamental property of astatine, the rarest
     atom on Earth', University of York, May 15, 2013.
     https://www.york.ac.uk/news-and-events/news/2013/research/astatine/

34. FASTEST COMPUTER

1   'Colossus, the world's first electronic computer', The National Museum
     of Computing.
     https://www.tnmoc.org/colossus

2   'Frontier supercomputer debuts as world's fastest, breaking exascale
     barrier', Oak Ridge National Laboratory, May 30, 2022.
     https://www.ornl.gov/news/frontier-supercomputer-debuts-worlds-
     fastest-breaking-exascale-barrier

35. REACHING FOR THE SKY

1   'Skyscrapers: The race to the top', Jonathan Glancy, BBC, January 5,
     2015.
     https://www.bbc.com/culture/article/20141216-skyscrapers-the-race-to-
     the-top

2   'A space elevator is possible with today's technology, researchers say',
     MIT Technology Review, September 12, 2019.
     https://www.technologyreview.com/2019/09/12/102622/a-space-elevator-
     is-possible-with-todays-technology-researchers-say-we-just-need-to-
     dangle/

36. MESMERISING MECHANISMS

1 'A Model of the Cosmos in the ancient Greek Antikythera Mechanism', T. Freeth, D. Higgon, A. Dacanalis et al., *Scientific Reports* 11 (5821), March 12, 2021.

https://www.nature.com/articles/s41598-021-84310-w

2 'The Babbage Difference Engine #2 at CHM', Computer History Museum, July 23, 2012.

https://www.youtube.com/watch?v=be1EM3gQkAY

37. WHOOSH

1 'Messerschmitt Me 163B-1a Komet', Royal Air Force Museum.

https://www.rafmuseum.org.uk/research/collections/messerschmitt-me-163b-1a-komet/

2 'How the Bell X-1 Ushered In the Supersonic Age', Jeff Macgregor, *Smithsonian*, October 2022.

https://www.smithsonianmag.com/smithsonian-institution/bell-x1-supersonic-flight-180980765/

38. THE MANY AND THE FEW

1 'Along with Humans, Who Else Is in the 7 Billion Club?', Bill Chappell, NPR, November 3, 2011.

https://www.npr.org/sections/thetwo-way/2011/11/03/141946751/along-with-humans-who-else-is-in-the-7-billion-club

39. DESCENT

1 'The daring journey inside the world's deepest cave', BBC, September 23, 2019.

https://www.bbc.com/reel/video/p07p40y7/the-daring-journey-inside-the-world-s-deepest-cave

2 'Project Mohole, 1958–1966', National Academy of Sciences.

http://www.nasonline.org/about-nas/history/archives/milestones-in-NAS-history/project-mohole.html

3 'The deepest hole we have ever dug', Mark Piesing, BBC, May 6, 2019.

https://www.bbc.com/future/article/20190503-the-deepest-hole-we-have-ever-dug

40. AGE MATTERS

1 'Hear the world's oldest instrument, the 50,000 year old Neanderthal flute', Sofia Rizzi, Classic FM, October 1, 2021.

https://www.classicfm.com/discover-music/instruments/flute/worlds-oldest-instrument-neanderthal-flute/

2 'Oldest piece of Earth discovered', Nadia Whitehead, *Science*, February 24, 2014.

https://www.science.org/content/article/oldest-piece-earth-discovered

3 'Hubble Finds Birth Certificate of Oldest Known Star', NASA, March 7, 2013.

https://www.nasa.gov/mission_pages/hubble/science/hd140283.html

41. SMALL

1 'Etruscan Shrew', Thai National Parks.

https://www.thainationalparks.com/species/etruscan-shrew

2 'What's the smallest thing in the Universe?', Jonathan Butterworth, TED-Ed, November 15, 2018.

https://www.youtube.com/watch?v=ehHoOYqAT_U

42. A SENSITIVE SUBJECT

1 'Just How Good Is Eagle Vision?', BBC Earth, March 24, 2023, YouTube video.

https://www.youtube.com/watch?v=A6H2ZdrKmzc

2 'Wax Moth Has Most Sensitive Ears in Insect World', Helen Fields, *Science*, May 7, 2013.

https://www.science.org/content/article/scienceshot-wax-moth-has-most-sensitive-ears-insect-world

3 'Magnetoreception in birds', Roswitha Wiltschko and Wolfgang Wiltschko, *Journal of the Royal Society*, September 4, 2019.

https://royalsocietypublishing.org/doi/10.1098/rsif.2019.0295

43. ERUPTION

1 'Volcanic eruption at Thera (Santorini)', Canadian Museum of History.

https://www.historymuseum.ca/cmc/exhibitions/civil/greece/gr1040e.html

2 'Volcanic Explosivity Index', National Parks Service.

https://www.nps.gov/subjects/volcanoes/volcanic-explosivity-index.htm

3 'Why the Yellowstone Supervolcano Could Be Huge', Smithsonian Channel, June 5, 2015, YouTube video.

https://www.youtube.com/watch?v=IMLo0E66O8A

44. THE METHUSELAH SYNDROME

1 'Netted whale hit by a lance a century ago', Erin Conroy, NBC News, June 12, 2007.

https://www.nbcnews.com/id/wbna19195624

2 'Greenland Sharks Live Hundreds of Years', Margaret Davis, Science Times, August 27, 2021.

https://www.sciencetimes.com/articles/33111/20210827/greenland-sharks-teach-humans-live-long.htm

3 'Scientists discover world's oldest clam, killing it in the process', Elizabeth Barber, *Christian Science Monitor*, November 5, 2013.

https://www.csmonitor.com/Science/2013/1115

4 'Methuselah, a Bristlecone Pine, Is Thought to Be the Oldest Living Organism on Earth', Robert Hudson Westover, U.S. Department of Agriculture, April 21, 2011.

https://www.usda.gov/media/blog/2011/04/21/methuselah-bristlecone-pine-thought-be-oldest-living-organism-earth

45. GREAT SURVIVORS

1 'In Ocean's Depths, Heat-Loving "Extremophile" Evolves a Strange Molecular Trick', Yale University, April 30, 2009.

https://news.yale.edu/2009/04/30/ocean-s-depths-heat-loving-extremophile-evolves-strange-molecular-trick

2 'Tardigrades: Nature's Great Survivors', Michael Marshall, *The Guardian*, March 20, 2021.

https://www.theguardian.com/science/2021/mar/20/tardigrades-natures-great-survivors

3 'Microbes found in natural asphalt lake', Lin Edwards, Phys.org, April 21, 2010.

https://phys.org/news/2010-04-microbes-natural-asphalt-lake.html

Image Credits

Page 6: Octobass of Montreal Symphony Orchestra
Courtesy of Wikimedia Commons: https://commons.wikimedia.org/wiki/File:Octobasse_Orchestre_Symphonique_de_Montr%C3%A9al_Eric_Chappell_1.jpg

Page 13: The University of Queensland's pitch drop experiment
Courtesy of Amada44, Wikimedia Commons: https://commons.wikimedia.org/wiki/File:University_of_Queensland_Pitch_drop_experiment-white_bg.jpg

Page 26: Microsoft's anechoic chamber
Courtesy of Ana Romero López, Wikimedia Commons: https://commons.wikimedia.org/wiki/File:Cámara_anecoica..jpg

Page 43: Inside the Joint European Torus
Courtesy of EUROfusion, Wikimedia Commons: https://commons.wikimedia.org/wiki/File:JET_vessel_internal_view_mascot.jpg

Page 48: Spheres on Gravity Probe B
Courtesy of NASA, Wikimedia commons: https://commons.wikimedia.org/wiki/File:Einstein_gyro_gravity_probe_b.jpg

Page 63: Silica aerogel
Courtesy of NASA/JPL-Caltech: https://solarsystem.nasa.gov/stardust/images/technology/aerogelhand.jpg

Page 70: Whispering Gallery of St Paul's Cathedral
Courtesy of Femtoquake, Wikimedia Commons: https://commons.wikimedia.org/wiki/File:St_Paul%27s_Cathedral_Whispering_Gallery.jpg

Page 77: *Sonic Wind 1* rocket sled and John Stapp
Image courtesy of US Air Combat Command: https://media.defense.gov/2014/Dec/08/2000982306/-1/-1/0/141208-F-CP123-003.JPG

Page 94: Laniakea
Courtesy of Andrew Z. Colvin, Wikimedia Commons: https://commons.wikimedia.org/wiki/File:07-Laniakea_(LofE07240).png

Page 110: Parker Solar Probe
Courtesy of NASA/Johns Hopkins APL/Steve Gribben: https://solarsystem.nasa.gov/missions/parker-solar-probe/in-depth/

Page 124: Greek black-figure pottery
Courtesy of Carole Radatto, Wikimedia Commons: https://commons.wikimedia.org/wiki/File:Athenian_black-figure_pottery_amphora,_5-6th_century_BC,_Theseus_slaying_the_Minotaur,_the_Cretan_monster,_Ashmolean_Museum_%2814338652154%29.jpg

Page 134: Enceladus
Courtesy of NASA/JPL/Space Science Institute: https://www.jpl.nasa.gov/edu/images/enceladus.jpg

Page 148: Porcupinefish
Courtesy of Mikkel Elbech, Wikimedia Commons: https://commons.wikimedia.org/wiki/File:Diodon_nicthemerus.jpg

Page 160: 17.5-ton truck supported by DELO glue
Courtesy of DELO: https://www.delo-adhesives.com/fileadmin/news/_testdateien/news/cross-industrial_information/gwr_image_4.jpg

Page 164: Titan arum
Courtesy of Sailing moose, Wikimedia Commons: https://commons.wikimedia.org/wiki/File:Amorphophallus_titanum_(corpse_flower)_-_2.jpg

Page 175: Giant soap bubble
Courtesy of Kazbeki, Wikimedia Commons: https://commons.wikimedia.org/wiki/File:Giant.bubble.jpg

Page 186: Peacock flounder
Courtesy of Brocken Inaglory, Wikimedia Commons: https://commons.wikimedia.org/wiki/File:Peacock_Flounder_Bothus_mancus_in_Kona.jpg

Page 201: Frontier supercomputer
Courtesy of Oak Ridge National Laboratory: https://www.ornl.gov/sites/default/files/2022-08/52281122090_6200529f58_o.jpg

Page 243: Bee hummingbird
Courtesy of SlvrHwk, Wikimedia Commons: https://commons.wikimedia.org/wiki/File:Mellisuga_helenae_Size_Comparison.svg

Page 256: Yellowstone
Courtesy of James St. John, Wikimedia Commons: https://commons.wikimedia.org/wiki/File:Grand_Prismatic_Spring_2013.jpg

Page 266: Tardigrade
Courtesy of Schokraie E, Warnken U, Hotz-Wagenblatt A, Grohme MA, Hengherr S, et al., Wikimedia Commons: https://commons.wikimedia.org/wiki/File:SEM_image_of_Milnesium_tardigradum_in_active_state_-_journal.pone.0045682.g001-2.png

Index

References to images are in *italics*.

© ALAN RICHARDSON

David Darling is a science writer, astronomer and musician. He is the author of more than fifty books, including the bestselling *Equations of Eternity*. Together with Agnijo Banerjee, he is the co-author of the *Weird Maths* trilogy and *The Biggest Number in the World*. He lives in Dundee, Scotland. Follow him on YouTube @drdaviddarling, Facebook @drdavid.darling and Twitter @drdaviddarling. Website: www.daviddarling.info.